Phycobiliproteins: Recent Developments and Future Applications

Vinod K. Kannaujiya
Shanthy Sundaram • Rajeshwar P. Sinha

Phycobiliproteins: Recent Developments and Future Applications

Springer

Vinod K. Kannaujiya
Centre for Biotechnology
University of Allahabad
Allahabad, Uttar Pradesh, India

Shanthy Sundaram
Centre for Biotechnology
University of Allahabad
Allahabad, Uttar Pradesh, India

Rajeshwar P. Sinha
Centre of Advanced Study in Botany
Banaras Hindu University
Varanasi, Uttar Pradesh, India

ISBN 978-981-10-6459-3 ISBN 978-981-10-6460-9 (eBook)
https://doi.org/10.1007/978-981-10-6460-9

Library of Congress Control Number: 2017961552

Printed on acid-free paper

This Springer imprint is published by Springer Nature
The registered company is Springer Nature Singapore Pte Ltd.
The registered company address is: 152 Beach Road, #21-01/04 Gateway East, Singapore 189721, Singapore

Preface

The biotechnological and industrial significance of phycobiliproteins (PBPs) for several commercial applications in the fields of biotechnology, biomedicine, pharmaceuticals, colorants, fluorescence agents, and food products is widely known. PBPs are major accessory light-harvesting antenna complexes, which exhibit incredible colors and play a vital role in the harvesting of light energy from the sun. The evolutionary diversity of ancient PBPs among cyanophyta, rodophyta, and cryptophyta facilitates research and development to understand hidden functional aspects of energy utilization. Future research will drive the field toward greater understanding of PBPs for proper utilization in various applications in biological science, including energy devices. This book encompasses topics of biotechnology, purification technology, molecular biology, production technology, biochemistry, stress biology, biomedicine, food technology, structural biology, genomics, and proteomics of PBPs.

Current advancements in chromatic adaptation and regulation of biosynthetic pathways of PBPs may enhance knowledge to improve bioengineering and biotechnological processes. Various kinds of photobioreactors are involved in large-scale cultivation and production of microalgae. Current advancements in industrial production of PBPs can overcome the costs of production, energy utilization, and imbalance of high-value food products for proper utilization in the biotechnology industry. Several techniques have been elaborated for extraction and purification of PBPs that are desired for quality products. The commercial and industrial applications of PBPs are widespread in many countries, in the forms of food color, pharmaceuticals, and fluorescent agents. Current research on PBPs provides new glimmers of hope for treatment of several diseases and improvement of human health.

This book presents state-of-the-art knowledge and a cohesive overview of current advancements in PBP development for biotechnological applications. Moreover, the book describes detailed techniques for purification, production, and commercial application of PBPs. In addition, it provides key notes regarding current advancements in research for understanding of concealed information about PBPs. This book is very helpful for biotechnology companies planning to venture into development of commercial biotechnological products.

We would like to express our profound appreciation for the generous support we have received from Springer. Dr. Vinod K. Kannaujiya is thankful to the University Grant Commission (UGC), New Delhi, India, for financial support in the form of a

Dr. D. S. Kothari Post-doctoral Research Grant (F.4-2/2006(BSR)/14-15/0526). Finally, we are also grateful to the Centre of Biotechnology, University of Allahabad, Allahabad, Uttar Pradesh, India, for providing the resources and facility for written materials.

Allahabad, Uttar Pradesh, India

Varanasi, Uttar Pradesh, India
April 22, 2017

Vinod K. Kannaujiya
Shanthy Sundaram
Rajeshwar P. Sinha

Contents

About the Authors

Vinod K. Kannaujiya completed his M.Sc. and Ph.D. in botany from the Centre of Advanced Study in Botany, Banaras Hindu University, Varanasi, India. He is currently working as Dr. D. S. Kothari postdoc fellow at the Centre of Biotechnology, Nehru Science Centre, University of Allahabad, Allahabad, India. He has been awarded with several prestigious national fellowships such as junior and senior research fellowship from the Council of Scientific and Industrial Research and national postdoctoral fellowship from DST, Govt. of India. He is also a life member of Indian Photobiology Society, India. He has published a number of papers in peer-reviewed journals and book chapters. He has also published several abstracts in national and international conferences. His main research interests include photobiological stress response, stability, and biotechnological and biomedical application of phycobiliproteins isolated from cyanobacteria inhabiting diverse habitats.

Shanthy Sundaram is professor of biotechnology and coordinator at the Centre of Biotechnology, Nehru Science Centre, University of Allahabad, Allahabad, India. She graduated with gold medal in M.Sc. microbiology at Bombay University, India, in 1986. She received doctorate degree in microbiology from Barkatullah University, Bhopal, India, in 1992. Dr. Sundaram has received many reputed awards and postdoctoral fellowships, including the Jawaharlal Nehru Memorial Trust UK Scholarship and Commonwealth Academic Staff Fellowship at the University of Warwick, UK. She has been working in the field of algae biotechnology for more than 22 years and published more than 100 papers in national and international reputed journals and conferences. She is a life member of various national and international scientific societies and reviewer of many national and international scientific journals. Her research activities are related to the development of advanced biotechnological tools in energy and environment area as biosensor, sustainable biofuel, cleaner technologies, etc.

Rajeshwar P. Sinha is a professor of molecular biology at the Centre of Advanced Study in Botany, Banaras Hindu University, Varanasi, India. He received his Ph.D. in biotechnology from Banaras Hindu University, Varanasi, India. He is a fellow of the Society for Applied Biotechnology, India, and recipient of the prestigious DAAD (Germany) fellowship. He has visited several countries such as Argentina, Austria, Belgium, Canada, China, Germany, Greece, France, Italy, Japan, Luxembourg, Norway, Poland, Republic of Korea, Spain, Switzerland, the Netherlands, and the UK, in the field of academics and research on one or the other scientific assignments. He has over 23 years of research and teaching experience. He has been working on the effects of UV radiation on aquatic ecosystems and has concentrated on the effects on DNA damage and repair, phycobiliproteins, mycosporine-like amino acids, and scytonemin induction in cyanobacteria and algae. He is a life member of various national and international scientific societies and an editorial board member of several national and international scientific journals. He has published over 200 original research papers, reviews, and book chapters and edited three books. He is having an h-index of 41 with over 6300 citations in reputed scientific journals including *Nature* and *Science*.

Introduction

1

Cyanobacteria are pioneer photosynthetic oxygen-evolving prokaryotes that appeared during the Precambrian era (2.8–3.5 billion year) and established a favorable condition for the evolution of current aerobic life (Fischer 2008). Cyanobacteria are phylogenetically a primitive group of Gram-negative prokaryotes having a cosmopolitan distribution ranging from hot springs to the Arctic and Antarctic regions (Stanier and Cohen-Bazire 1977). These are the most abundant nitrogen-fixing microorganisms commonly reside in low land soil of the rice paddy fields (Roger and Kulasooriya 1980; Häder et al. 2015). There are certain cyanobacteria that make symbiotic association with plants. Symbiont of *Anabaena* and *Azolla* plays an indispensable role in nitrogen fixation (Vaishampayan et al. 2001). The promising role of cyanobacteria as nitrogen fixer in rice field is well documented (Sinha and Häder 1996; Sinha et al. 2001; Tirkey and Adhikary 2005). Proliferations of cyanobacteria in rice fields are due to anaerobic nitrogen fixation through the heterocyst as well as some vegetative cells of non-heterocystous cyanobacteria (Prasanna and Kaushik 1994). They also play a major role in successional processes and global photosynthetic biomass production and nutrient cycling. Thus, cyanobacteria are important as primary producer in both terrestrial and aquatic ecosystems.

Cyanobacteria are dependent on solar energy for photosynthesis and nitrogen fixation. Light-harvesting antenna complexes (phycobiliproteins) are major component in cyanobacteria and red algae that exhibit incredible colors and play a vital role in the harvesting of light energy from the sun (Fig. 1.1).

Phycobiliproteins (PBPs) efficiently absorb sunlight in wavelengths ranging from 480 to 660 nm and transiently transfer to the main reaction center of chlorophyll *a* (Sun et al. 2003). These can be classified into three main groups such as C-phycoerythrin (C-PE), C-phycocyanin (C-PC), and C-allophycocyanin (C-APC) having absorption maxima between 540–570, 610–620, and 650–655 nm, respectively (Santiago-Santos et al. 2004; Bermejo et al. 2003) (Fig. 1.2).

The color of PBPs originates mainly from covalently bound prosthetic groups containing open chain 1–4 tetrapyrrole chromophore bearing A, B, C, and D pyrrole ring (Grossman et al. 1993; Padyana et al. 2001). Phycobiliproteins (PBPs) are

V.K. Kannaujiya et al., *Phycobiliproteins: Recent Developments and Future Applications*, https://doi.org/10.1007/978-981-10-6460-9_1

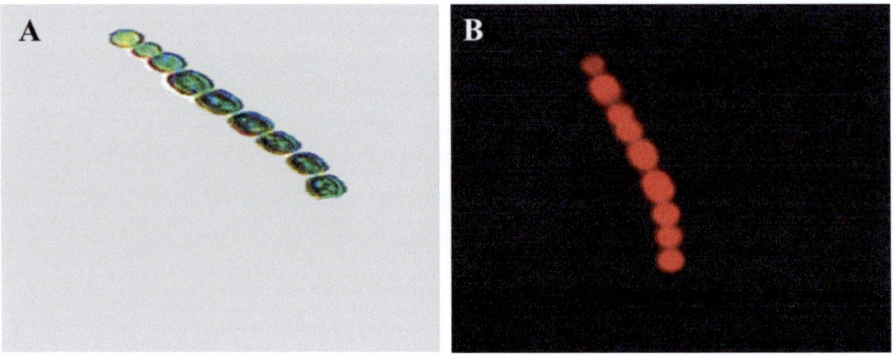

Fig. 1.1 Light microscopic image (**a**) and fluorescence image (**b**) of vegetative cells of *Nostoc* sp. HKAR-2, showing PBPs

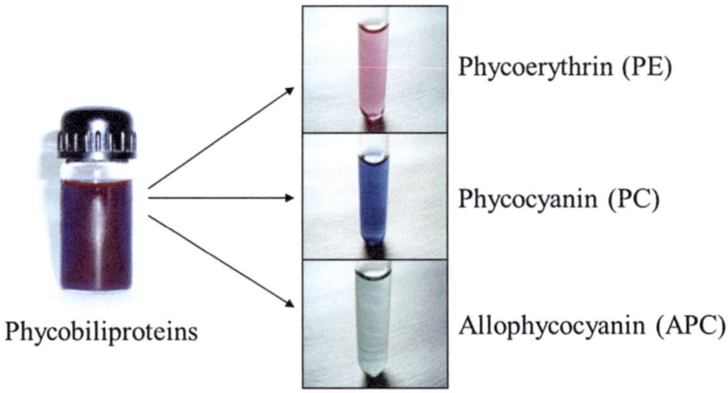

Fig. 1.2 Color compositions of phycobiliproteins

brilliantly colored water-soluble and autofluorescence accessory light-harvesting macromolecules organized in supramolecular complexes, called phycobilisomes (PBSs) on photosynthetic apparatus in cyanobacteria, red algae, and cryptomonads. PBPs are light-harvesting multi-protein complex which are arranged in rows and coupled to photosystem II (PS II) unit on the external surface of the thylakoid membranes (Glazer 1985).

Photoautotrophic is the largest outdoor method of PC production by cultures of cyanobacteria in open sunlight. In outdoor open raceways, the areal productivities of dry biomass and PC in cultures of *Arthrospira platensis* and *Anabaena* sp. have reached values of 14–23.5 and 0.82–1.32 g m^{-2} day^{-1}, respectively (Graverholt and Eriksen 2007). Modern food industry leads to an increase in the production of cheaper, healthier, and more convenient food products without any harm to human body. Therefore, consumer's demand for more natural food products, having health benefits, has increased over the years. PBPs are one of the most promising food products that are being used efficiently by people as additive and marketed as food and cosmetic colorant in Japan. However, the consumption of blue foods (PC) has been limited, probably due to unawareness and unproductive industrial sector. Apart from nutritional value of PBPs, it stimulates the immune defense system and possesses antioxidant, anti-inflammatory, antiviral, anticancer, and cholesterol-lowering effects. Interestingly, PC has also been shown to significantly inhibit cell proliferation; thus it can be used as anticarcinogenic agent. PBPs are currently being used as natural colorants for food such as chewing gum and dairy products. PC is more stable than indigo and gardenia and emits a bright blue fluorescent color in jelly gum, soft candies, fermented milk products, ice creams, soft drinks, desserts, sweet cake decoration, milk shakes, and cosmetics (Richa et al. 2011). The advancement in purification technology may increase the application of individual PC, PE, and APC in biomedical and biotechnological sciences. Application of acetone/ammonium sulfate precipitation, gel filtration, and ion exchange chromatography techniques has been found to be effective for the enrichment and purification of PBPs (Kannaujiya and Sinha 2016). Recently, two-phase aqueous extraction followed by ion exchange chromatography has resulted in extremely pure PBPs with the highest purity index. The future of wide application of PBPs depends on commercialization and improvements in bioprocess engineering for making high quality products. The fluorescence properties of PBPs are tremendously used in the detection of biomolecules in various fields of biotechnology. PBPs are widely used in immunology laboratories, as they can serve as labels for antibodies, receptors, and other biological molecules. PBPs conjugated to immunoglobulins, protein A, and avidin were developed into excellent reagents for two-color fluorescence analysis of single cell using fluorescence-activated cell sorter (FACS) and also used in histochemistry, fluorescence microscopy, and fluorescence immunoassays. Recent studies have shown that PC has health promising and a broad range of potential pharmaceutical applications. The pharmacological property attributed by PC, PE, and APC includes antioxidant, anti-inflammatory, neuroprotective, and hepatoprotective activity. The pigment has also been shown to have protective effects in human pancreatic cells and against arthritis in rats by attenuating oxidative stress. Such findings about the PC show the potential benefits in the prevention of many pathological disorders associated with oxidative stress and inflammation. The anticancer potential of PC isolated from cyanobacteria is well known. Daily ingestion of a small dosage of PC could maintain or accelerate lymphocytic functions to prevent malignancy such as cancer or to inhibit its growth or recurrence. The role of PC as antiviral, antitumor, antimicrobial, anti-HIV, and a food additive has also been well established. PC is also used for the treatment of diseases such as Alzheimer's and Parkinson's and prevents constipation, pancreatitis, cataract, skin cancers, and oral and degenerative diseases (Sekar and Chandramohan 2008). Apoptosis and anti-inflammatory effects have recently been confirmed (Fig. 1.3).

The use of PC in photodynamic therapy is an emerging biomedical application that has recently been confirmed. During recent years, global attention has been focused on cyanobacteria and red algae for their potential applications in food, feed, fuel, fertilizer, biopolymers, natural colorants, vitamins, toxins, enzymes, pharmaceuticals, pharmacological fluorescent probes, and pollution abatement (Sekar and

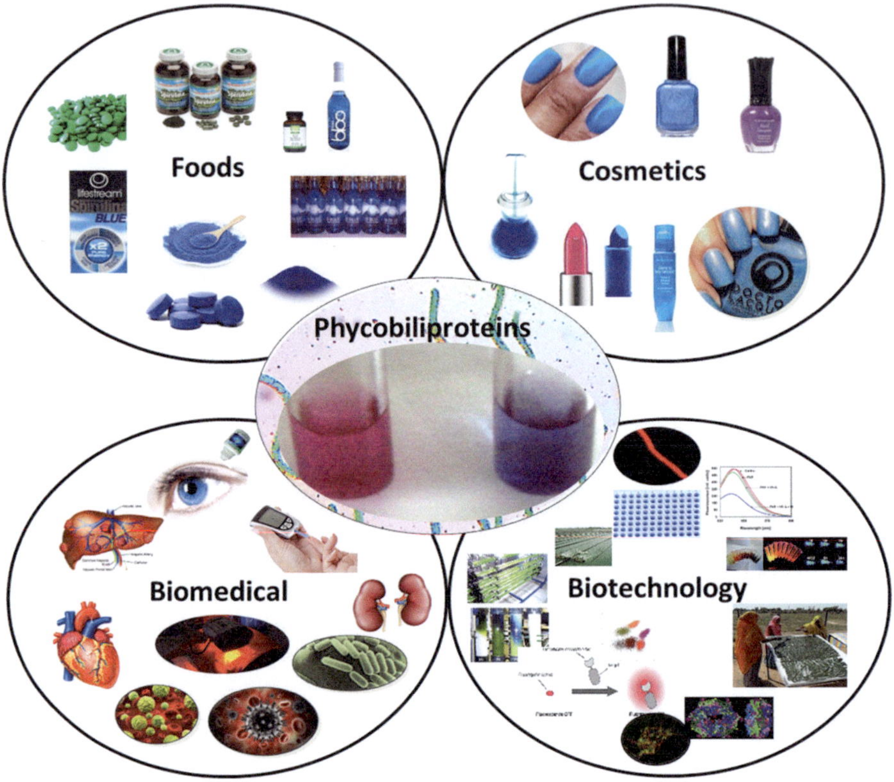

Fig. 1.3 Diverse applications of phycobiliproteins

Chandramohan 2008). In the corporate sector, more than 15 companies are involved in the production and selling of PBPs.

The aim of this book is to outline the basic understanding in the diversity of PBPs with structure and function for better utilization. We have mainly focused on the recent developments in molecular structure and function, mass production, purification, and utilization of PBPs in the field of biotechnology, pharmacology, and food applications. The probable role in therapeutic science will also be discussed in detail. This book may play indispensable role in improvement in knowledge of fundamental and various applications of PBPs in biotechnology and biomedical sciences.

References

Bermejo R, Acién FG, Ibáñez MJ, Fernández JM, Molina E, Alvarez-Pez JM (2003) Preparative purification of B-phycoerythrin from the microalga *Porphyridium cruentum* by expanded-bed adsorption chromatography. J Chromatogr B 790:317–325

Fischer WF (2008) Biogeochemistry: life before the rise of oxygen. Nature 455:1051–1052

Glazer AN (1985) Light harvesting by phycobilisomes. Annu Rev Biophys Chem 14:47–77

Graverholt OS, Eriksen NT (2007) Heterotrophic high-cell density fed-batch and continuous-flow cultures of *Galdieria sulphuraria* and production of phycocyanin. Appl Microbiol Biotechnol 77:69–75

Grossman A, Schaefer MR, Chiang GG, Collier JL (1993) The phycobilisomes, a light-harvesting complex responsive to environmental conditions. Microbiol Rev 57:725–749

Häder D-P, Williamson CE, Wängberg S, Rautio M, Rose KC, Gao K, Helbling EW, Sinha RP, Worrest R (2015) Effects of UV radiation on aquatic ecosystems and interactions with other environmental factors. Photochem Photobiol Sci 14:108–126

Kannaujiya VK, Sinha RP (2016) An efficient method for the separation and purification of phycobiliproteins from a rice-field cyanobacterium *Nostoc* sp. strain HKAR-11. Chromatographia 79:335–343

Padyana AK, Bhat VB, Madyastha KM, Rajashankar KR, Ramakumar S (2001) Crystal structure of a light-harvesting protein C-phycocyanin from *Spirulina platensis*. Biochem Biophys Res Commun 282:893–898

Prasanna R, Kaushik BD (1994) Physiological and molecular genetic aspects of nitrogen fixation in non-heterocystous cyanobacteria. Indian J Exp Biol 32:248–251

Richa KVK, Kesheri M, Singh G, Sinha RP (2011) Biotechnological potentials of phycobiliproteins. Int J Pharm Bio Sci 2:446–454

Roger PA, Kulasooriya SA (1980) Blue-green algae and rice. The International Rice Research Institute, Los Baños

Santiago-Santos MC, Ponce-Noyola T, Olvera-Ramírez R, Ortega-López J, Cañizares-Villanueva RO (2004) Extraction and purification of phycocyanin from *Calothrix* sp. Process Biochem 39:2047–2052

Sekar S, Chandramohan M (2008) Phycobiliproteins as a commodity: trends in applied research, patents and commercialization. J Appl Phycol 20:113–136

Sinha RP, Häder D-P (1996) Photobiology and ecophysiology of rice field cyanobacteria. Photochem Photobiol 64:887–896

Sinha RP, Klisch M, Gröniger A, Häder D-P (2001) Responses of aquatic algae and cyanobacteria to solar UV-B. Plant Ecol 154:221–236

Stanier RY, Cohen-Bazire G (1977) Phototrophic prokaryotes: the cyanobacteria. Annu Rev Microbiol 31:225–274

Sun L, Shumei W, Chen L, Gong X (2003) Promising fluorescent probes from phycobiliproteins. IEEE J Sel Top Quant 9:177–188

Tirkey J, Adhikary SP (2005) Cyanobacteria in biological soil crusts of India. Curr Sci 89:515–521

Vaishampayan A, Sinha RP, Häder D-P, Dey T, Gupta AK, Bhan U, Rao AL (2001) Cyanobacterial biofertilizers in rice agriculture. Bot Rev 67:453–516

Evolution of Phycobiliproteins

2

2.1 Introduction

Cyanobacteria are pioneer for oxygenic photosynthesis on Earth's surface for survivability of life. Phycobiliproteins (PBPs) are brightly colored accessory light-harvesting complex integrated in photosynthetic apparatus of cyanobacteria and certain branch of macroalgae such as red algae, cryptomonads, and rare group of glaucophytes (Grossman et al. 1993; Sidler 1994; Sinha et al. 1995; Apt et al. 1995). During evolutionary development, eukaryotic photosynthetic organism shows distinct variability in photosystem for efficient utilization of solar spectrum. The interrelationship of photosynthesis machinery between prokaryote and eukaryote may be developed by endosymbiotic mechanism with cyanobacteria. Indeed, most of the accessory light-harvesting units are embedded in photosynthetic units in various autotrophic organisms which reflect the abundance in ecological and adaptive significance in various environmental conditions. In cyanobacteria, thylakoid membrane is embedded with pigment such as photosystem (PSI, PSII) and outer surface surrounded by colored PBPs (Colyer et al. 2005). The formation of functional PBP core assembly involved colored PBPs (85%) and colorless linker polypeptides (15%) (Parsiegla et al. 2012). All the subunits of PBPs are water-soluble, highly fluorescent, and brilliantly colored proteins that are categorized mainly into three groups such as phycocyanin (PC), phycoerythrin (PE), and allophycocyanin (APC) (Bermejo et al. 2003; Santiago-Santos et al. 2004; Sinha et al. 1995). The basic structural arrangements of core units of PBPs are made up of α- and β-chains in cyanobacteria whereas γ chain found in some red algae. Most of the core chains of PBPs are evolved from same ancestor (Thomas and Passaquet 1999). Primarily, α- and β-chains are key components for functional integrity of PC, PE, and APC. PBPs consist of diverse group of amino acid constituents embedded with linear tetrapyrrole chromophore at cysteine positions (Kannaujiya et al. 2014). The molecular weights of α subunits (12–20 kDa) are comparatively lower than β subunits (15–22 kDa) in PBPs (Kannaujiya et al. 2014; Sinha et al. 1995). In genomic analysis, nucleotides may influence amino acid residue variability and functional codon

V.K. Kannaujiya et al., *Phycobiliproteins: Recent Developments and Future Applications*, https://doi.org/10.1007/978-981-10-6460-9_2

usage of PC, PE, and APC that may take part in evolution (Kannaujiya et al. 2014). The large diversity of PBPs pigmentation is to facilitate through gene exchanges in lineages that maintain the diversity of micro-marine environment. The prominent role and molecular mechanism of PBPs in photosynthesis have been widely examined. Nevertheless, evolution of the photosynthetic system is not completely understood. In this chapter, we describe certain mechanism of evolution of PBPs and focus on gene flow between plastids and PBPs.

2.2 Origin of Photosynthesis

It is well understood that the development of photosynthetic light-harvesting organelles in eukaryote is exclusive products of endosymbiosis mechanism that occurs between photosynthetic eukaryote and cyanobacterium. The cascade of gene analysis that encoded mitochondria (Burger et al. 1999), plastid (Delwiche and Palmer 1997; Douglas and Penny 1999), and nucleus genome (Baldauf et al. 2000; Moreira et al. 2000) of higher plants clearly signifies the endosymbiotic mechanism with photosynthetic prokaryotes (Fig. 2.1). The advancement of plastid is evolved by

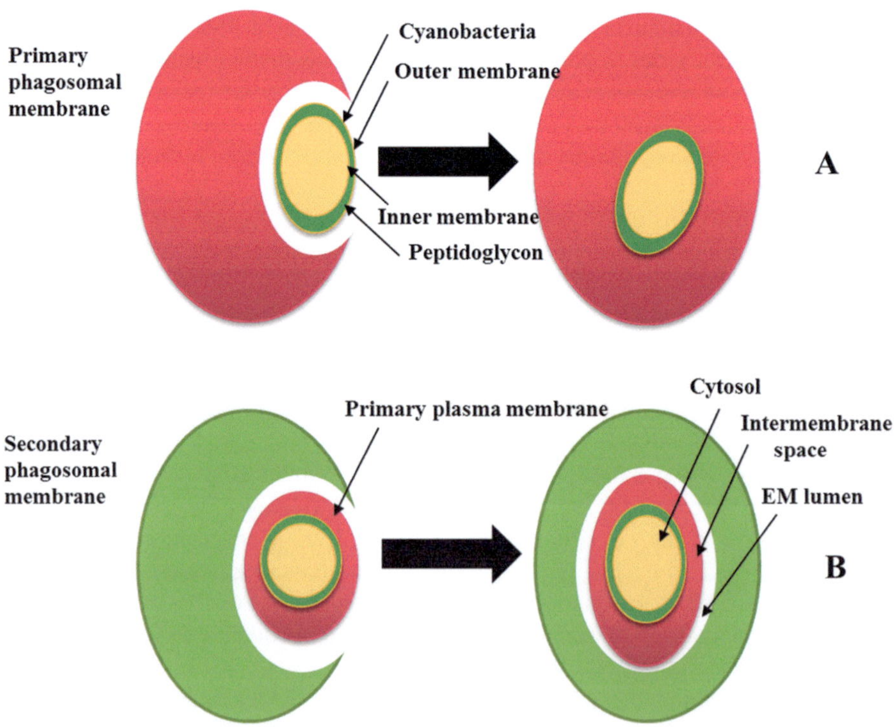

Fig. 2.1 Evolution of photosynthetic system in prokaryotes and eukaryotes from cyanobacteria through primary (**a**) and secondary (**b**) endosymbiosis mechanism (Modified from Archibald and Keeling 2002)

secondary endosymbiosis that differentiates the organism into red and green lineage (Büchel 2015).

The secondary endosymbiosis has changed the classification of photosynthetic organism from endosymbionts of chlorarachniophytes and euglenophytes to chromalveolates that includes cryptomonads, haptophytes, heterokontophytes, diatoms, and dinoflagellates (Archibald and Keeling 2002). Structurally, the primary endosymbiotic plastids are enclosed by double membranes similar to Gram-negative cell wall envelope. Glaucocystophytes have distinct peptidoglycan cell wall that shows remarkable evidence for evolution of plastids from cyanobacteria (Archibald and Keeling 2002). Secondary (or complex) plastids are present in certain algae and structurally categorized by further addition of membranes that includes three membranes (euglenids and dinoflagellates) and four membranes (heterokonts, haptophytes, apicomplexa, cryptomonads, and chlorarachniophytes) (Archibald and Keeling 2002).

2.3 Evolution of Phycobilisomes

2.3.1 Sequence Alignment and Amino Acid Variability

The diversity of PBPs and its accessory linker polypeptides originated from the common ancestor has been well recognized. Each of the PBP subunits has a composition of nine α-helical domains separated by asymmetrical loops which include E-X and Y-A helices found in both subunits for structural protein-protein interactions and stabilization of αβ heterodimer subunits (Apt et al. 1995). The trimeric aggregation of PBPs $(\alpha\beta)_3$ is made up by reunion of identical αβ monomers and two trimers associated together to form hexameric $(\alpha\beta)_6$ face-to-face arrangement. Linker polypeptides are associated with central cavity by interaction with rod-rod, rod-core, and core-core of the trimers and hexamers of PBPs. There is a diverse kind of linker polypeptides associated with various kind of PBP hexamer. It has been well recognized that there is a similarity in the composition of PBPs subunits among various cyanobacteria which indicates its evolution from common ancestor proteins (Glazer et al. 1976; Glazer 1980; Kannaujiya et al. 2014). There are several conserved amino acid residues which interact with chromophores for structural stability of PBPs. Structurally, conserved aspartate amino acid at 91 position interacts directly with chromophore by 88 position of same residue (Duerring et al. 1990; Schirmer et al. 1986) and plays a critical role for maintaining interaction between chromophores. Moreover, β PC and β PEC have also interacted with conserved aspartate residues at 40 positions in 3D architecture. α subunit of PBPs has quite fewer Cys and Met residue, while Ala, Gly, and Leu exist in dominant residues. However, β subunit has more Asp and Val residue as compared to other. The occurrence of His and Trp is found negligible in number in both α and β subunits (Kannaujiya et al. 2014). The multiple sequence alignment shows a supplementary binding site of chromophores (PEII) at specific location of aspartic acid residue. Moreover, TEA may act as terminal energy acceptor to PSII which regulated different attachment sites of chromophores (Glazer 1989). The β subunits are enriched with conserved hydrophobic residue (alanine) that interacted with couple of arginine amino acids at 80 and 81 positions (Schirmer et al. 1986). However, α subunits are interacted by hydrophilic residues such as threonine (or lysine) at the same position of chromophores. Thus, α subunit chromophore environment may create probability of unidirectional energy transfer to form β subunits (Apt et al. 1995). Evolutionary evidence suggests that 3D complex of globin protein may act as ancestor for development of PBPs (Pastore and Lesk 1990; Schirmer et al. 1985).

2.3.2 Phylogenetic Analysis

Apt et al. (1995) advocated that evolution of PBPs may influence in the direction of increasing in absorption at shorter wavelengths; unfortunately this evolutionary hypothesis has not been adopted practically. The ancestry tree indicates close relationship between α and β subunits in terms of physical and energetic interactions in distinct and symmetrical lines that are required for specific coevolutionary adaptations of each class of PBPs (Fig. 2.2). The hypothesis of evolutionary adaptive changes in dN/dS values can strengthen variation analysis of known structural interaction of proteins (Zhao and Qin 2006).

Moreover, an elevated dN/dS ratio levels with positive evolutionary selection in respect to chromophore-binding regions of various kind of PBPs lineages were

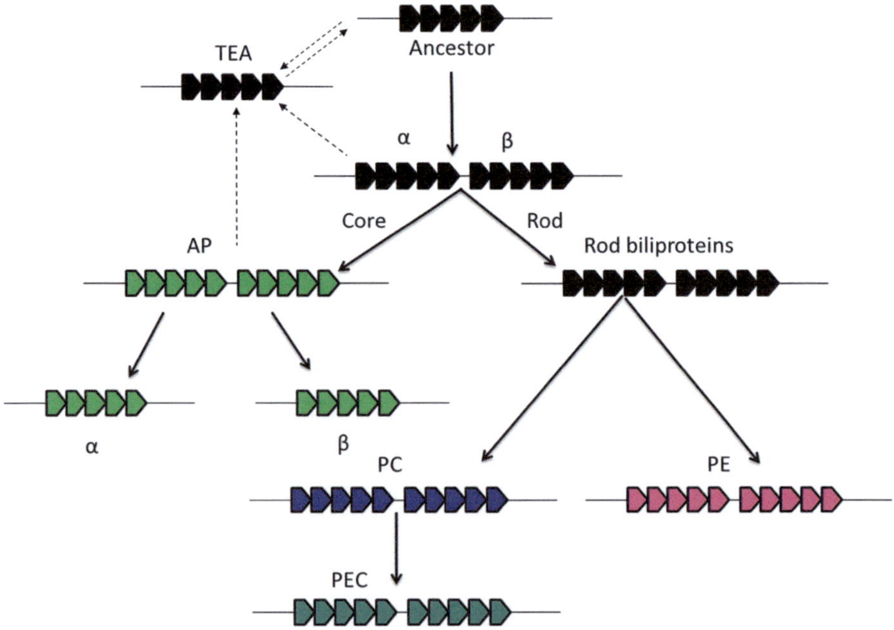

Fig. 2.2 Evolution of genes encoding α and β subunits of core and rod phycobiliproteins in cyanobacteria (Adapted and modified from Apt et al. 1995)

isolated from different habitats. The association of residues around chromophores was also identified by elevated levels of the same ratio in most of conserved position of PBPs that help in the transfer of harvested light energy (Sidler 1994; Apt et al. 1995). Domain-related perturbation experiment at specific site could cause alteration in spectroscopic property which further confirms the spatial requirement necessary for a transmission of resonance energy transfer between chromophores of PBPs (Martinez-Oyanedel et al. 2004).

The phenotypic analysis of PE is not mapped on phylogenetic analysis based on their housekeeping genes of organism (Toledo et al. 1999Fuller et al. 2003; Six et al. 2007; Everroad 2007). The nucleotide analyses of genes in PE collected from various freshwater cyanobacteria, marine cyanobacteria (*Synechococcus* sp.), and red algae have revealed that the evolutionary flow of PBS genes from core genome may be unspecified events of horizontal gene transfer (Six et al. 2007; Everroad 2007). Recently, Jasser et al. (2011) have proposed horizontal gene transfer mechanism found in cluster of PC and PE in the phylogenetic tree of *cpcBA*-IGS that explain the phenotypic distribution of PBPs. Everroad and Wood (2006) illustrated that marine *Synechococcus* 5.1 isolates have shared a common ancestry with *Cyanobium* sp.; however it is not possible to determine the relationship in development from direct ancestor of PBPs (PE). Similarly, phylogenetic characteristics of PE containing *Prochlorococcus* lineages are quite similar to PUB containing strains *Synechococcus* 5.1 (Six et al. 2007; Dufresne et al. 2008). Doust et al. (2004) have developed a structural as well as functional model for PE545 isolated from cryptophyte. The spectral differences of PE545 were mainly alteration in chemical constituents between the chromophores and binding relationship to local residue environment. The gene sequences of PE isolated from *Prochlorococcus* show evolutionarily isolated group by their differential molecular selection of PE in various taxonomic category (Ting et al. 2001). More specifically, CpeBA (PE) protein sequences obtained from strains *Synechococcus* and *Cyanobium* sp. (PUB-lacking phenotype) are exclusively categorized together in sister monophyletic origin of *cpeBA* gene sequences. The phylogenetic relationship shows distinct evidence of progressive molecular evolution of photosynthetic units by mechanism of gene duplication. Picocyanobacteria (marine) are evolutionary model organism for adaptive phenotypic evolution in nature (Everroad and Wood 2012). The consequences of gene duplication of ancestral PBPs gave rise to dual independent lines such as core (APC) and rod (PC and PEC) subunits with exclusion of linker polypeptides (Fig. 2.3).

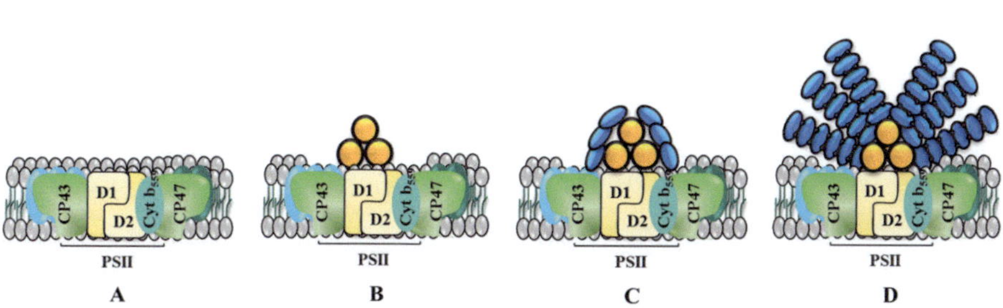

Fig. 2.3 Structural development and evolution of phycobiliproteins

Table 2.1 Characteristics of linker polypeptides found in microalgae

Proteins	Symbol	Amino acids	MW (kDa)	P*i*	Annotation	Gene
CpeC	PE L_R	285–294	31.8–33.1	9.6	PE-associated linker	*cpeC*
CpeD	PE L_R	249–255	27.9–28.4	8.2–8.6	PE-associated linker	*cpeD*
CpeE	PE L_R	244–254	27.1–28.4	9.7	PE-associated linker	*cpeE*
CpcC	PC L_R	219–291	24.8–32.6	9.5–9.6	PC-associated linker	*cpcC*
CpcD	PC L_R	70–87	7.8–9.9	9.8–10.5	Rod capping linker	*cpcD*
CpcG	L_{RC}	231–279	26.8–31.9	9.3–9.6	Rod-core linker	*cpcG*
CpcH	L_{RC}	271–273	30.4–30.8	8.8–9.7	Rod-core linker	*cpcH*
CpcI	L_{RC}	288	32.7	8.9	Rod-core linker	*cpcI*
ApcC	L_C	66–69	7.7–7.8	10.9–11.4	APC-associated linker	*apcC*
ApcE	L_{CM}	683–1155	76.5–129.8	9.5–9.7	Core-membrane linker	*apcE*

Adapted from Liu et al. (2005)

In addition, PECs form a separate group that includes combination with phycobiliviolin chromophores which efficiently absorb green light in environmental condition (Bryant 1982). There were silent changes in substitution of synonymous sites of PBPs which have no impacts on three-dimensional structure of protein. However, nonsynonymous substitutions are assembled together on various domains of chromophores after the gene duplication (X and Y). The nonsynonymous substitution shows a significant correlation and promotes coevolution of PBPs. Thus, the helical native domain contains two subunits (X and Y) of PBPs that were considered for structural and functional importance of chromophores of monomeric subunits.

2.3.3 Linker Polypeptide Gene Diversity

The linker polypeptides are colorless and integrated with PBP subunits by inter-/intramolecular interaction (Liu et al. 2005). These linker polypeptides have wide range of molecular mass (8–120 kDa) that play important role for stabilization of structural and functional property of PBP complex (Table 2.1). Moreover, linker polypeptides also facilitate assembly of PBP complex and relative interaction of chromophores for modulation of light absorption that promotes unidirectional transfer of energy from PBPs to photosystem (Glazer1989; Bryant 1991; Grossman et al. 1993). Linker polypeptides are categorized into four groups: (a) core linkers (LRC) associated with peripheral rods of core PBS; (b) rod linkers such as L_R10, L_R33, and L_R35 that interacted with PC and PE substructures' rod-rod interaction; (c) LC8 small core linker polypeptides that are interlinked with trimeric APC in central cavity; and (d) core-membrane linker polypeptides such as LCM99 which interacted with PBS core to PSII organization in membrane that play major role for energy regulation (Glazer1989; Capuano et al. 1991; Sidler 1994).

Interestingly, most of the linker polypeptides are colorless while certain core linker polypeptides (LCM) associated with PCB-ApcE and PE (I,II) γ subunit chromophores with characteristic color (Scheer and Zhao 2008). Due to brief knowledge about tertiary structures of linker polypeptides and its interaction with PBPs, mapping of selected sites on PBPs is not possible. Moreover, previous studies revealed that six motifs of LR and LRC sequences show quite similarity which correlated the functional importance of PBPs (Glauser et al. 1992; Sidler 1994). Moreover, linker polypeptides occupy internal cavities of rod and core disks and are incorporated in stabilization, assembly of rod-core, as well as directionality transfer of energy in PBPs (Adir and Lerner 2003).

2.4 GC Regulation and Codons

The fast progress in genome annotation and various bioinformatics tools has untied various research avenues to reconnoiter hidden molecular biology in pigment system of cyanobacteria. In contrast to bacterial species, the investigation about variability of nucleotide and amino acid codons from various nucleotide/protein compositions of cyanobacteria is now easy due to progress of full genome sequencing. Sueoka (1961) defines the correlation bias between GC constituents and amino acid composition in bacterial genome. GC constituents are broad and ranges from 22.5% to 72% in bacterial genome (Fig. 2.4).

Recently, GC content ranges showed more variation from 16.6% to 74.9% in whole bacterial genome (Lightfield et al. 2011). The variability of synonymous codons has not been changed by amino acid that nullifies the bias in species by preferential amino acid codon selection (Sharp et al. 1995; Agashe et al. 2013). Similarly, few residues such as Ala, Gly, Pro, and Arg are preferentially expressed by GC-rich codons, although nucleotide bias still exists in base composition (Lobry 1997; Singer and Hickey 2000; Bharanidharan et al. 2004). It was found that heterologous gene expression may promote bias in codon usage (Plotkin and Kudla 2011). Apart from bacteria, total genome of cyanobacterial strains shows lower variations (35–56%) in GC nucleotide (Li and Watanabe 2002). The dinucleotide parameters GC and related amino acid codons are play distinguished role in evolution of PBPs (Kannaujiya et al. 2014) (Fig. 2.5).

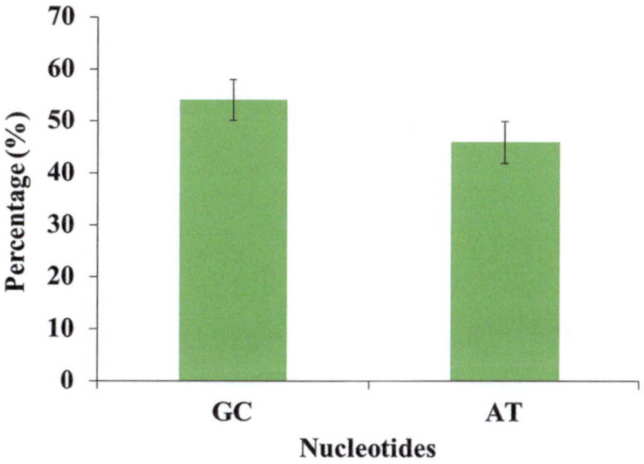

Fig. 2.4 Nucleotide percentage in phycobiliproteins of cyanobacteria

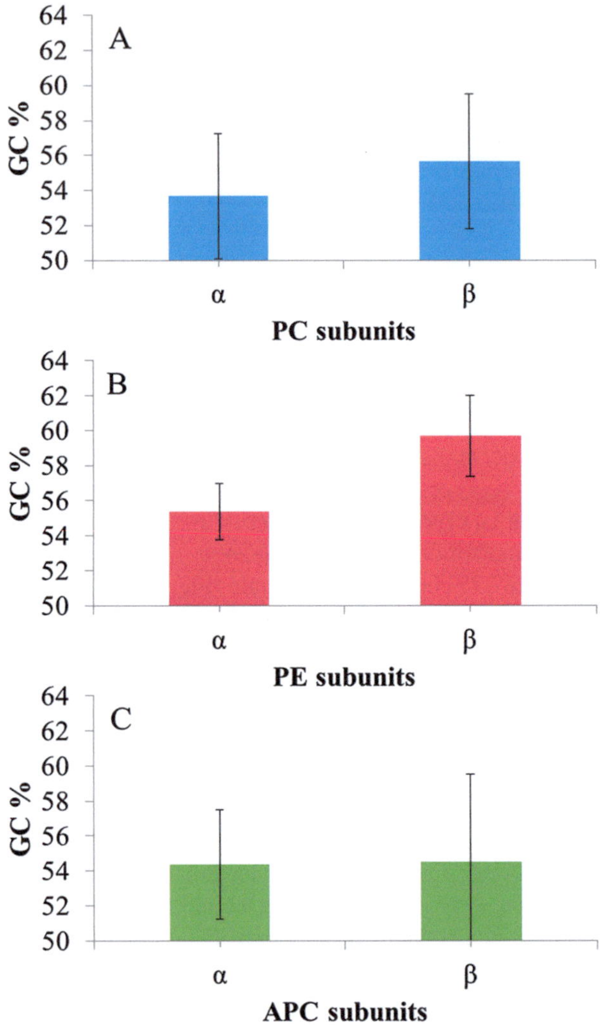

Fig. 2.5 Average nucleotide composition (*GC*) in PC (**a**), PE (**b**), and APC (**c**) subunits of cyanobacteria

Apart from GC, AT also exists from 40% to 51% that may play distinguished role in evolution (Li and Watanabe 2002). Except of cyanobacteria, bacteria have more wide GC variability that ranges from 16.6% to 74.9% (Lightfield et al. 2011). However, prokaryotic organism may maintain codon accuracy in mutational bias in nucleotide and translational selection by preference for selection of codons (Shah and Gilchrist 2011; Plotkin and Kudla 2011; Xu et al. 2013). Furthermore, amino acids that oriented protein expression in prokaryotes are also influenced by fluctuation in environment (Subramaniam et al. 2013). In addition, third position of GC nucleotide in *Synechococcus* sp. plays a distinguished role in habitat-induced translational selection of codons for amino acids (Nair et al. 2013).

The levels of GC play the most important role in nucleotide variation in codons, and amino acids may influence mechanism of mutation and translational selection

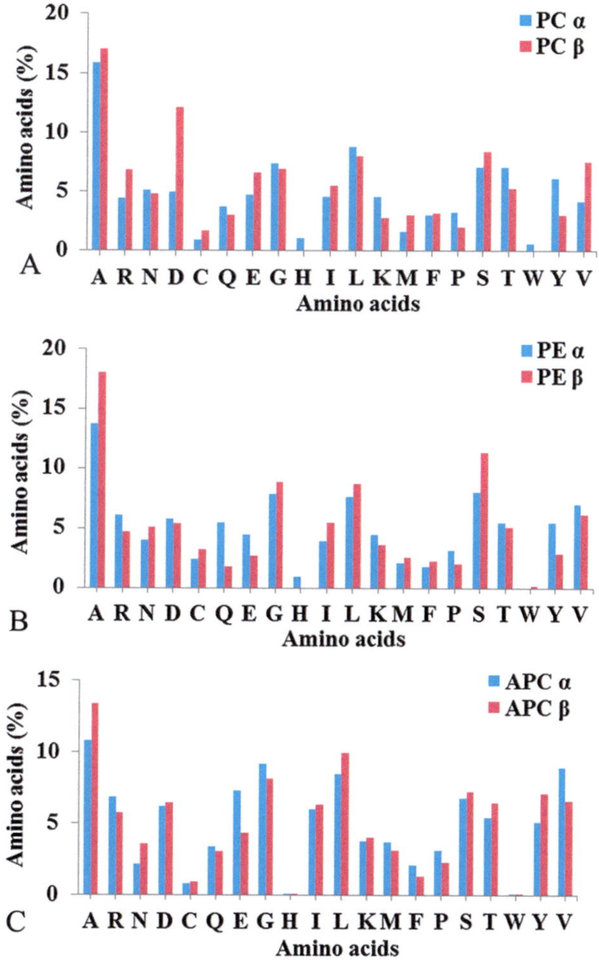

Fig. 2.6 The relative amino acid composition of PC (**a**), PE (**b**), and APC (**c**) subunits in phycobiliproteins

of codons (Lynch et al. 2008; Yu et al. 2012). Moreover, correlation and regression analysis between GC levels (G1C1, G2C2, G3C3) provides hypothesis that indicates inconsistency which originated from mutational and translational selection pressure (Nair et al. 2013; Sueoka 1988). Similarly, the variability of G+C levels at GC1 to GC3 codon levels was found in bacterial genome (Gupta and Ghosh 2003). Interestingly, it was found that GC levels have significant correlation with certain amino acid residue (Ala, Arg, Gly, Pro), while negative correlation was shown with other amino acid residue (Asp, Ile, Lys, Met, Phe, Tyr) (Banerjee et al. 2005). In aerobic prokaryotes, certain amino acid residues such as Cys, His, Met, Phe, Trp, and Tyr are expressed in low quantity as compared to anaerobic prokaryotes due to radical-induced oxidation of residue (Berlett and Stadtman 1997; Naya et al. 2002). Cyanobacteria show distinguished variability in α and β subunits of PC, PE, and APC (Fig. 2.6).

2.5 Conclusion

Evolutionary studies have focused particularly on closely related gene lineage to address the roles of selective pressures for adaptation of domains. The genes of PBPs may be evolved by elevated rates of nonsynonymous substitutions in functional genome. More specifically, codon substitution analyses of specific residue of amino acids in nucleotide of PBPs gene segment will provide a key evidence for evolutionary phenomena in cyanobacterial proteins. The nucleotide variation in GC has more specific tool for codon expression. The current progress in complete genome sequencing of cyanobacteria may help in understanding of PBPs as well as plastid evolution over mutational pressure in the coming future.

References

Adir N, Lerner N (2003) The crystal structure of a novel unmethylated form of C-phycocyanin, a possible connector between cores and rods in phycobilisomes. J Biol Chem 278:25926–25932

Agashe D, Gomez NCM, Drummond DA, Marx CJ (2013) Good codons, bad transcript: large reductions in gene expression and fitness arising from synonymous mutations in a key enzyme. Mol Biol Evol 30:549–560

Apt KE, Collier JL, Grossman AR (1995) Evolution of the phycobiliproteins. J Mol Biol 248:79–96

Archibald JM, Keeling PJ (2002) Recycled plastids: a green movement in eukaryotic evolution. Trends Genet 18:577–584

Baldauf SL, Roger AJ, Wenk-Siefert I, Doolittle WF (2000) A kingdom-level phylogeny of eukaryotes based on combined protein data. Science 290:972–977

Banerjee T, Gupta SK, Ghosh TC (2005) Role of mutational bias and natural selection on genome-wide nucleotide bias in prokaryotic organisms. Biosystems 81:11–18

Berlett BS, Stadtman ER (1997) Protein oxidation in aging, disease and oxidative stress. J Biol Chem 272:313–316

Bermejo R, Acién FG, Ibáñez MJ, Fernández JM, Molina E, Alvarez-Pez JM (2003) Preparative purification of B-phycoerythrin from the microalga *Porphyridium cruentum* by expanded-bed adsorption chromatography. J Chromatogr B 790:317–325

Bharanidharan D, Bhargavi GR, Uthanumallian K, Gautham N (2004) Correlations between nucleotide frequencies and amino acid composition in 115 bacterial species. Biochem Biophys Res Commun 315:1097–1103

Bryant DA (1982) Phycoerythrocyanin and phycoerythrin: properties and occurrence in cyanobacteria. J Gen Microbiol 128:835–844

Bryant DA (1991) Cyanobacteria phycobilisomes; progress towards complete structural and functional analysis via molecular genetics. In: Bogorad L, Vasil IK (eds) Cell culture and somatic cell genetics of plants, molecular biology of plastids and mitochondria. Academic Press Inc, San Diego, pp 257–300

Büchel C (2015) Evolution and function of light harvesting proteins. J Plant Physiol 172:62–75

Burger G, Saint-Louis D, Gray MW, Lang BF (1999) Complete sequence of the mitochondrial DNA of the red alga *Porphyra purpurea*. Cyanobacterial introns and shared ancestry of red and green algae. Plant Cell 11:1675–1694

Capuano V, Braux AS, Tandeau de Marsac N, Houmard J (1991) The anchor polypeptide of cyanobacterial phycobilisomes. Molecular characterization of the *Synechococcus* sp. PCC 6301 *apce* gene. J Biol Chem 266:7239–7247

Colyer CL, Kinkade CS, Viskari PJ, Landers JP (2005) Analysis of cyanobacterial pigments and proteins by electrophoretic and chromatographic methods. Anal Bioanal Chem 382:559–569

Delwiche CF, Palmer JD (1997) The origin of plastids and their spread via secondary endosymbiosis. In: Bhattacharya D (ed) Origins of algae and their plastids. Springer-Verlag, Wien, pp 53–86

Douglas SE, Penny SL (1999) The plastid genome from the cryptomonad alga, *Guillardia theta*: complete sequence and conserved synteny groups confirm its common ancestry with red algae. J Mol Evol 48:236–244

Doust AB, Marai CN, Harrop SJ, Wilk KE, Curmi PM, Scholes GD (2004) Developing a structure-function model for the cryptophyte phycoerythrin 545 using ultrahigh resolution crystallography and ultrafast laser spectroscopy. J Mol Biol 344:135–153

Duerring M, Huber R, Bode W, Ruembeli R, Zuber H (1990) Refined three-dimensional structure of phycoerythrocyanin from the cyanobacterium *Mastigocladus laminosus* at 2.7Å. J Mol Biol 211:633–644

Dufresne A, Ostrowski M, Scanlan DJ, Garczarek L, Mazard S, Palenik BP, Paulsen IT, de Marsac NT, Wincker P, Dossat C, Ferriera S, Johnson J, Post AF, Hess WR, Partensky F (2008) Unraveling the genomic mosaic of a ubiquitous genus of marine cyanobacteria. Genome Biol 9:R90

Everroad RC (2007) Diversification of marine picocyanobacteria: the ecology and evolution of spectral phenotype and phycoerythrin. The University of Oregon, Eugene

Everroad RC, Wood AM (2006) Comparative molecular evolution of newly discovered picocyanobacterial strains reveals a phylogeneticaly informative variable region of b-phycoerythrin. J Phycol 42:1300–1311

Everroad RC, Wood AM (2012) Phycoerythrin evolution and diversification of spectral phenotype in marine *Synechococcus* and related picocyanobacteria. Mol Phylogenet Evol 64:381–392

Fuller NJ, Marie D, Partensky F, Vaulot D, Post AF, Scanlan DJ (2003) Clade-specific 16S ribosomal DNA oligonucleotides reveal the predominance of a single marine *Synechococcus* clade throughout a stratified water column in the Red Sea. Appl Environ Microbiol 69:2430–2443

Glauser M, Bryant DA, Frank G, Wehrli E, Rusconi SS, Sidler W, Zuber H (1992) Phycobilisome structure in the cyanobacteria *Mastigocladus laminosus* and *Anabaena* sp. PCC 7120. Eur J Biochem 205:907–915

Glazer AN (1980) Structure and evolution of photosynthetic accessory pigment systems with special reference to phycobiliproteins. In: Sigman DS, MAB B (eds) The evolution of protein structure and function. Academic, New York, pp 221–244

Glazer AN (1989) Light guides. Directional energy transfer in a photosynthetic antenna. J Biol Chem 264:1–4

Glazer AN, Apell GS, Hixson CS, Bryant DA, Rimon S, Brown DM (1976) Biliproteins of cyanobacteria and rhodophyta: homologous family of photosynthetic accessory pigments. Proc Natl Acad Sci U S A 73:428–431

Grossman A, Schaefer MR, Chiang GG, Collier JL (1993) The phycobilisomes, a light-harvesting complex responsive to environmental conditions. Microbiol Rev 57:725–749

Gupta SK, Ghosh TC (2003) Reinvestigation on the causes of genomic GC variation between the orthologous genes of *Mycobacterium tuberculosis* and *Mycobacterium leprae*. Biochem Biophys Res Commun 303:65–68

Jasser I, Królicka A, Karnkowska-Ishikawa A (2011) A novel phylogenetic clade of picocyanobacteria from the Mazurian lakes (Poland) reflects the early ontogeny of glacial lakes. FEMS Microbiol Ecol 75:89–98

Kannaujiya VK, Rastogi RP, Sinha RP (2014) GC constituents and relative codon expressed amino acid composition in cyanobacterial phycobiliproteins. Gene 546:162–171

Li R, Watanabe MM (2002) DNA base composition of planktonic species of *Anabaena* (Cyanobacteria) and its taxonomic value. J Gen Appl Microbiol 48:77–82

Lightfield J, Fram NR, Ely B (2011) Across bacterial phyla, distantly-related genomes with similar genomic GC content have similar patterns of amino acid usage. PLoS One 6:e17677

Liu L-N, Chena X-L, Zhanga Y-Z, Zhoub B-C (2005) Characterization, structure and function of linker polypeptides in phycobilisomes of cyanobacteria and red algae: an overview. Biochim Biophys Acta 1708:133–142

Lobry JR (1997) Influence of genomic G+C content on average amino-acid composition of proteins from 59 bacterial species. Gene 205:309–316

Lynch M, Sung W, Morris K, Coffey N, Landry CR, Dopman EB, Dickinson WJ, Okamoto K, Kulkarni S, Hartl DL, Thomas WK (2008) A genome wide view of the spectrum of spontaneous mutations in yeast. Proc Natl Acad Sci U S A 105:9272–9277

Martinez-Oyanedel J, Contreras-Martel C, Bruna C, Bunster M (2004) Structural-functional analysis of the oligomeric protein R-phycoerythrin. Biol Res 37:733–745

Moreira D, Guyader HL, Philippe H (2000) The origin of red algae and the evolution of chloroplasts. Nature 405:69–72

Nair RR, Nandhini MB, Sethuraman T, Doss G (2013) Mutational pressure dictates synonymous codon usage in freshwater unicellular α-cyanobacterial descendant *Paulinella chromatophora* and β-cyanobacterium *Synechococcus elongatus* PCC 6301. Springer Plus 2:492

Naya H, Romero H, Zavala A, Alvarez B, Musto H (2002) Aerobiosis increases the genomic guanine plus cytosine content (GC %) in prokaryotes. J Mol Evol 55:260–264

Parsiegla G, Shrestha B, Carrière F, Vertes A (2012) Direct analysis of phycobilisomal antenna proteins and metabolites in small cyanobacterial populations by laser ablation electrospray ionization mass spectrometry. Anal Chem 84:34–38

Pastore A, Lesk AM (1990) Comparison of the structure of globins and phycocyanins: evidence for evolutionary relationship. Proteins Struct Funct Genet 8:133–155

Plotkin JB, Kudla G (2011) Synonymous but not the same: the causes and consequences of codon bias. Nat Rev Genet 12:32–42

Santiago-Santos MC, Ponce-Noyola T, Olvera-Ramírez R, Ortega-López J, Cañizares-Villanueva RO (2004) Extraction and purification of phycocyanin from *Calothrix* sp. Process Biochem 39:2047–2052

Scheer H, Zhao KH (2008) Biliprotein maturation: the chromophore attachment. Mol Microbiol 68:263–276

Schirmer T, Bode W, Huber R, Sidler W, Zuber H (1985) X-ray crystallographic structure of the light-harvesting biliprotein C-phycocyanin from the thermophilic cyanobacterium *Mastigocladus laminosus* and its resemblance to globin structures. J Mol Biol 184:257–277

Schirmer T, Huber R, Schneider M, Bode W, Miller M, Hackert ML (1986) Crystal structure analysis and refinement at 2.5Å of hexameric C-phycocyanin from the cyanobacterium *Agmenellum quadruplicatum*. The molecular model and its implications for light harvesting. J Mol Biol 188:651–676

Shah P, Gilchrist MA (2011) Explaining complex codon usage patterns with selection for translational efficiency, mutation bias, and genetic drift. Proc Natl Acad Sci U S A 108:10231–10236

Sharp PM, Averof M, Lloyd AT, Matassi G, Peden JF (1995) DNA sequence evolution: the sounds of silence. Philos Trans R Soc B 349:241–247

Sidler WA (1994) Phycobilisome and phycobiliprotein structures. In: Bryant DA (ed) The molecular biology of cyanobacteria. Kluwer Academic Publication, Dordrecht, pp 139–216

Singer GAC, Hickey DA (2000) Nucleotide bias causes a genome wide bias in the amino acid composition of proteins. Mol Biol Evol 17:1581–1588

Sinha RP, Lebert M, Kumar A, Kumar HD, Häder D-P (1995) Spectroscopic and biochemical analyses of UV effect on phycobiliprotein of *Anabena* sp. and *Nostoc carmium*. Bot Acta 108:87–92

Six C, Thomas JC, Garczarek L, Ostrowski M, Dufresne A, Blot N, Scanlan DJ, Partensky F (2007) Diversity and evolution of phycobilisomes in marine *Synechococcus* spp.: a comparative genomics study. Genome Biol 8:R259

Subramaniam AR, Pan T, Cluzel P (2013) Environmental perturbations lift the degeneracy of the genetic code to regulate protein levels in bacteria. Proc Natl Acad Sci U S A 110:2419–2424

Sueoka N (1961) Correlation between base composition of deoxyribonucleic acid and amino acid composition and protein. Proc Natl Acad Sci U S A 47:1141–1149

Sueoka N (1988) Directional mutation pressure and neutral molecular evolution. Proc Natl Acad Sci U S A 85:2653–2657

Thomas JC, Passaquet C (1999) Characterization of a phycoerythrin without α-subunits from a unicellular red alga. J Biol Chem 274:2472–2482

Ting CS, Rocap G, King J, Chisholm SW (2001) Phycobiliprotein genes of the marine photosynthetic prokaryote *Prochlorococcus*: evidence for rapid evolution of genetic heterogeneity. Microbiology 211:3171–3182

Toledo G, Palenik B, Brahamsha B (1999) Swimming marine *Synechococcus* strains with widely different photosynthetic pigment ratios form a monophyletic group. Appl Environ Microbiol 65:5247–5251

Xu Y, Ma P, Shah P, Rokas A, Liu Y, Johnson CH (2013) Non-optimal codon usage is a mechanism to achieve circadian clock conditionality. Nature 495:116–120

Yu T, Li J, Yang Y, Qi L, Chen B, Zhao F, Bao Q, Wu J (2012) Codon usage patterns and adaptive evolution of marine unicellular cyanobacteria *Synechococcus* and *Prochlorococcus*. Mol Phylogenet Evol 62:206–213

Zhao F, Qin S (2006) Evolutionary analysis of phycobiliproteins: implications for their structural and functional relationships. J Mol Evol 63:330–340

Structural and Functional Significance of Phycobiliproteins

3

3.1 Introduction

Phycobilisomes (PBS) is a giant accessory light-harvesting complex made up of brilliantly colored biliproteins (BPs) that assemble with two sub-complexes, a central core, and array of two to five cylinders containing six to eight rods (Adir et al. 2006). Structurally, PBSs are supramolecular complexes with different components of BPs with broad molecular mass (5–30 MDa) (Watanabe and Ikeuchi 2013). The BPs are integrally bound to the protein scaffold containing fluorescence cofactor collectively called phycobiliproteins (PBPs). The stability of PBPs is dependent on colorless linker polypeptides (Liu et al. 2005). PBPs are brilliantly colored, water-soluble, and highly fluorescent accessory light-harvesting proteins located on photosynthetic membrane in cyanobacteria, red algae, glaucophytes, and cryptomonads for efficient harvesting of solar photonic energy for photosynthesis (Schirmer et al. 1985). On the basis of color and absorption properties, they are mainly categorized into three main groups such as phycoerythrin (PE), phycocyanin (PC), and allophycocyanin (APC). In addition, each subunit of PBPs consists of heterodimer α and β subunits which consisted of linear tetrapyrrole chromophores (one to four chains with A, B, C, and D ring) covalently attached to different cysteine residues of biliproteins (Sidler 1994). The open-chain tetrapyrrole chromophores are able to capture photonic energy of solar spectrum and transiently transfer to photosystem for utilization in photosynthesis. Regarding the pathway of energy transfer, it may start form terminal PE to PC and ultimately reach to APC and finally deliver to photosystem for carbon storage (Johnson et al. 2014). The optimization of light-harvesting properties and its energy transfer mechanism may be regulated by transient change in ratio of PC and PE during drastic changes in environment. About 20% of PC and APC are located in the periphery and core region, respectively, of the photosynthetic membrane (Singh and Montgomery 2013). In addition, PE is located at the distal region in array of PBs and plays an important role in the adaptation against drastic changes in photosynthetically active radiation (Rodríguez et al. 1991). A number of colorless linker polypeptides stabilize the overall structure of the PBS complex and

V.K. Kannaujiya et al., *Phycobiliproteins: Recent Developments and Future Applications*, https://doi.org/10.1007/978-981-10-6460-9_3

serve as structural elements, involved in the biosynthesis and stabilization of PBPs (Ana et al. 2009). Spectroscopic properties of the PBP subunits are due to their delocalization of conjugated pi–pi electrons transition and conjugated double bonds inside the pyrrole ring (Sun et al. 2009).

The aim of this chapter is to outline the structural and functional integrity of PBPs and elucidate the detail interrelationship between bilin chromophore and its fluorescence property for efficient transfer of excitation energy during photosynthesis.

3.2 Occurrence and Diversity

Cyanobacteria are photoautotrophic prokaryotes that evolved during the Precambrian era (2.8–3.5 billion year ago) and play a pioneering role for establishment of an oxygen-rich inhabitable environment for the evolution of existing aerobic life (Fischer 2008). Cyanobacteria are an unusual group of primitive photosynthetic Gram-negative prokaryotes having a cosmopolitan distribution in both aquatic and terrestrial ecosystems ranging from marine to freshwater and hot springs to Arctic and Antarctic regions (Stanier and Cohen-Bazire 1977; Rastogi et al. 2012). They are also significant constituent of marine ecosystem and account for 40% of oceanic primary productivity of an estimated biomass (Sinha and Häder 1996). Being cosmopolitan in distribution, cyanobacteria are exposed to different types of stresses in their natural environments. Cyanobacteria are a natural source for the production of various biologically important macromolecules such as lipids/fatty acids, bioactive compounds, proteins, enzymes, pigments, etc., which have immense pharmaceutical and nutraceutical values (Rastogi and Sinha 2009). Cryptophytes (genus *Cryptomonas*) are single-celled photosynthetic eukaryotic algae found abundantly in freshwater as well as marine water. The existence of algae is located at the bottom of freshwater/marine water, where light irradiation is very weak. The evolution of algae is assumed to be originated from cyanobacteria and red algae after endosymbiosis.

3.2.1 Occurrence of Phycobiliproteins in Organism

Cyanobacteria and red algae are dependent on photonic energy of solar spectrum for carbon storage and nitrogen fixation. Light-harvesting antenna complexes are major component in cyanobacteria and red algae that exhibit incredible colors and play a vital role in the harvesting of light energy from the sun. This may be the main reason for survival of cyanobacteria and red algae due to presence of major accessory light-harvesting complexes where green algae and higher plant face inability to survive in their exclusively habitats using chlorophyll-proteins for harvesting energy from sun (Fig. 3.1). Thus, the presence of PBPs in red algae (eukaryotic) occupies a significant position by inherent evolution of oxygen-evolving cyanobacteria followed by predecessor of cryptophytes in eukaryotic organism (Allen et al. 2011; Price et al.

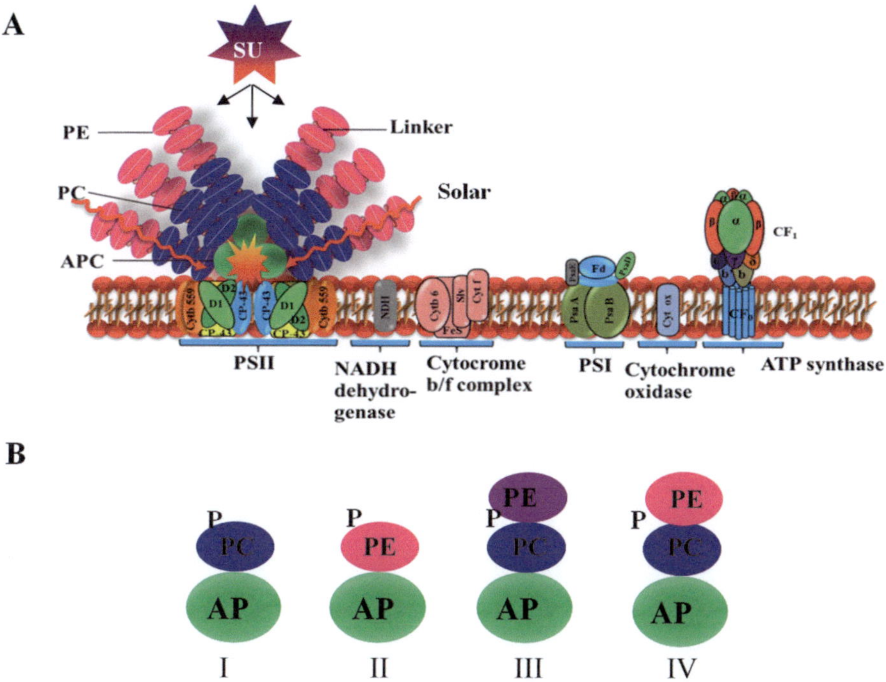

Fig. 3.1 Schematic structure of PBSs on thylakoid membrane depicted as rod subunits PC (phycocyanin) and PE (phycoerythrin) and core subunits AP/APC (allophycocyanin) with linker polypeptides (**a**) and a four-basic-PBP composition found in cyanobacteria/red algae (**b**) (For details, see text)

2012). R-PE was reported in red algae *Porphyridium cruentum* which profusely grows under intertidal zone of marine habitats; thus it may further confirmed the survivability of organism similar to ancient prokaryotic cyanobacteria (Gantt et al. 2003; Su et al. 2010); Sun and Wang 2011). The red algae are able to grow abundantly in deep water column under low light irradiation due to the presence of high content of PE which efficiently absorb solar radiation in the range of 450–570 nm. However, cyanobacteria show a broad range of absorption wavelengths ranging from 480 to 660 nm as compared to red algae (Sun et al. 2003).

Cryptophytes have accessory light-harvesting properties homologous to PBPs found in red algae. Interestingly, the cryptophyte BPs are found on dense granular matrix of thylakoid instead of stromal surface of thylakoid (Spear-Bernstein and Miller 1989). The aggregation of heterodimeric structure has changed into trimeric form. There are found two cryptophyte biliproteins such as phycoerythrin 545 (PE545) and phycocyanin 645(PC645) isolated from *Rhodomonas* CS24 and *Chroomonas* CCMP270, respectively (Doust et al. 2006). The chlorophyll-*f* containing cyanobacterium, *Halomicronema hongdechloris*, is grown under monochromatic far-red light (730 nm) and efficiently harvests light energy with the help of

PBPs (Chen et al. 2012). Surprisingly, the PBPs isolated from *Halomicronema hongdechloris* have the ability to absorb light beyond >700 nm under far-red region. The structural analysis of PBPs has revealed a new strategy for light adaptation, where loss of PC-containing rods and remodeling of the core subunits (APC) with novel far-red light absorption property (Chen et al. 2012; Li et al. 2016). Moreover, the molecular understanding of their remodeled PBPs at various levels may contribute to a new finding and overcome the challenges of agricultural production of proteins (Ort et al. 2015).

3.2.2 Fluorescence and Spectroscopic Properties

These PBPs are composed of two types of proteins: colored phycobiliproteins (PBPs), which absorb and transmit light energy to the photosynthetic reaction centers, and nonpigmented linker proteins, which are organized into the PBPs and modulate their light absorption property (Glazer 1989). Cyanobacterial and red algae group of phycobiliproteins are classified into four forms such as phycocyanin (PC; λ_{max} = 610–625 nm), phycoerythrin (PE; λ_{max} = 540–570 nm), phycoerythrocyanin (PEC; λ_{max} = 560–600 nm), and allophycocyanin (APC; λ_{max} = 650–660 nm) (Sidler 1994; Santiago-Santos et al. 2004; Bermejo et al. 2003). In addition, on basis of the species, some of the PEs are divided into three main groups: B-phycoerythrin (B-PE; λ_{max} = 546–565 nm), C-phycoerythrin (C-PE; (λ_{max} = 565 nm), and R-phycoerythrin (R-PE; (λ_{max} = 545–565 nm) (Bermejo et al. 2003; Sun et al. 2003) (Table 3.1).

B-PE is extracted and purified from unicellular red alga *P. cruentum*, R-PE found in most of red macroalgae, certain group of CU-PE observed in some marine cyanobacteria (Six et al. 2007), and C-PE occurred in most of cyanobacteria (Fig. 3.2).

The color of the PBPs is originated mainly from the covalently bound prosthetic group. These prosthetic groups are open-chain tetrapyrrole chromophores bearing A, B, C, and D ring named as phycobilins. They are blue-colored phycocyanobilin (PCB), purple-colored phycobiliviolin (PVB), red-colored phycoerythrobilin (PEB), and yellow-colored phycourobilin (PUB).

3.3 Ultra Structure

Structurally, PBPs exist in hexameric form $(\alpha\beta)_6$, which are composed of six α and six β subunits, forming a flattened disc (~11 nm diameter) with a central cavity. The hexameric PBP complexes are comprised of two trimers arranged face to face with an internal cavity. It is generally believed that the cavity, with a diameter of approximately 3.5 nm, is occupied by linker polypeptides derived from the PBP crystals (Stadnichuk et al. 2015a).

Table 3.1 Spectroscopic and fluorescence properties of phycobiliproteins (For details see the references cited in the text)

Phycobiliproteins	Absorption maxima (nm)	Absorption color (visible spectrum)	Fluorescence emission maxima (nm)	Fluorescence color (ultraviolet radiation)	Distribution
C-Phycoerythrin (C-PE)	540–565	Pink, red	568–575	Yellow orange	Cyanobacteria
C-Phycocyanin (C-PC)	610–620	Purple, violet	630–650	Strong red	Cyanobacteria
Phycoerythrocyanin (PEC)	570–590	Purple, light red	600–610	Orange	Cyanobacteria
R-Phycoerythrin (R-PE)	562–565	Red	570–575	Yellow orange	Red algae
R-Phycocyanin (R-PC)	617–620	Purple, violet	635–638	Strong red	Red algae/ cyanobacteria
Allophycocyanin (APC)	650–655	Violet, gray	660–665	Weak red	Red algae/ cyanobacteria

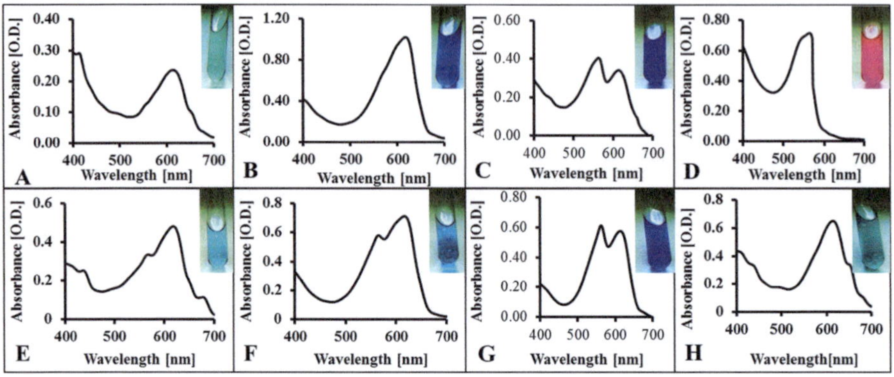

Fig. 3.2 Absorption spectrum of PBPs isolated from different cyanobacteria. (**a**) *Cyanothese* sp. HKAR-1, (**b**) *Anabaena* sp. BT-2, (**c**) *Nostoc* sp. HKAR-2, (**d**) *Anabaena circinalis*, (**e**) *Westiellopsis* sp., (**f**) *Rivularia* sp. HKAR-4, (**g**) *Nostoc* sp. HKAR-11, and (**h**) *Scytonema* sp. HKAR-3. Phycocyanin, blue in color; Phycoerythrin, pink/red in color; and Allophycocyanin, bluish green in color

3.3.1 Architecture of Phycobiliproteins

PBP complex is a very large protein complex, which can be easily visualized by use of electron microscopy (EM). It has been reported that some PBS models have interlink attachment of rods and core by using EM. Certain types of morphological structure of PBS have been described in cyanobacteria and red algae: (1) hemi-discoidal, (2) hemi-ellipsoidal, (3) bundle shaped, and (4) block shaped (Bryant et al. 1979). Hemi-discoidal PBS attached to the stromal side of the thylakoid membrane has a dimension of about 70 nm along the base, 30–50 nm in height, and 14–17 nm in width (Rosinski et al. 1981). PBPs are typically hemi-discoidal with diameters in the range of 32–70 nm (Grossman et al. 1993). In hemi-discoidal PBPs, there are generally six rods and three cylinders in their core (Fig. 3.3).

The bundle-shaped PBS is found in cyanobacterium, *Gloeobacter violaceus* (Guglielmi et al. 1981), and rod-like PBS is found *Acaryochloris marina*, a chlorophyll *d*-containing cyanobacterium (Marquardt et al. 1997). Subfractionation of PBS bundle-shaped model (*Gloeobacter violaceus*) provides a molecular orientation of linker polypeptides such as CpeG and CpcJ (Koyama et al. 2006). Recently, single-particle analysis has revealed the detail architecture of rods and core protein including linker polypeptides of PBS (Arteni et al. 2009; Yi et al. 2005; Watanabe and Ikeuchi 2013).

3.3.2 Stoichiometric Arrangement of Subunits

The self-assembly of all PBPs is initiated by the docking of two subunits abbreviated by α and β subunits with close homologous up to 25–40 % on the amino acid sequence at individual level, but the sequences become highly homologous on the

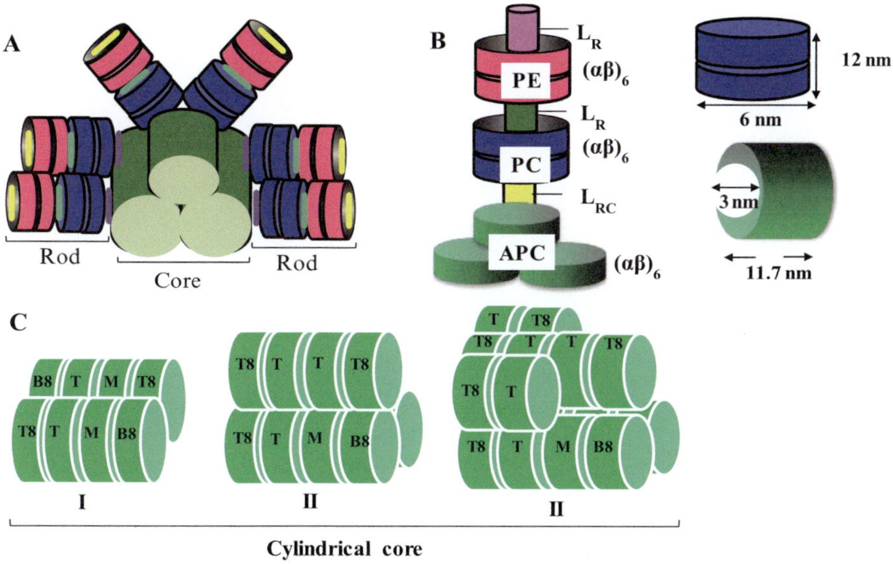

Fig. 3.3 Schematic representation of the three-dimensional intact PBS showing the PC/PE rod arrayed around the APC core in a parallel orientation (**a**), PBP dimension and associated linker polypeptides (**b**), and schematic arrangement of cylindrical core (allophycocyanin) (**c**). *I*, two identical asymmetric cylinders are arranged in anti-parallel manner, which is made up of four allophycocyanin trimer discs (T8/T/M/B8); *II*, tricylindrical core consists of two types of cylinder (T8/T/M/B8 and T8/T/T/T8) of allophycocyanin core; and *III*, pentacylindrical core contains two extra asymmetric half cylinders (only two trimer discs, T-T8) beside the symmetric cylinder as shown in figure, oriented in anti-parallel manner to each other. T8-contains α and β subunits of allophycocyanin and linker peptides (LC 8.9); T-disc contains α and β subunit of allophycocyanin without any linker peptide; and M-disc contains of α, β, and variant β (16 kDa) subunits of allophycocyanin and linker peptide (LCM 72–127). B8 contains α, variant α, and β subunits of allophycocyanin and linker peptide (LC 8.9) (Adapted and modified from Singh et al. 2015)

level of tertiary structure (McGregor et al. 2008; Adir et al. 2006; Apt et al. 1995). This heterodimer of α and β monomer is collectively recognized as the basic building block of monomeric structure of PBPs, which further assembles together into trimeric discs $(\alpha\beta)_3$. Some experimental data evident that the trimeric structures of biliproteins are further matured by attachment of chromophore and further assemble into $(\alpha\beta)_6$ hexamers with involvement of rod substructures and unpigmented linker polypeptides (Fig. 3.4).

Dissociated subunits typically have less intense color and mostly differentiated from native trimeric units into monomer subunits. Native trimeric and hexameric units have different molecular weight in different PBPs. The assembly of hexamers is more stable and highly functional for harvesting light energy from the sun. However, the trimeric form of many PCs has shown an interesting assembly of rod structures without involvement of linker polypeptides (Adir 2005). In core cylinder, four trimers of APC biliproteins assemble together into core cylinders. Subsequently, two to five core cylinders also assemble into the core substructure with involvement

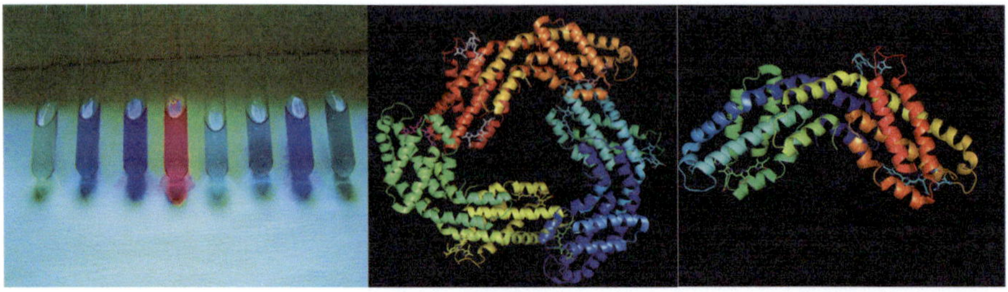

Fig. 3.4 Three-dimensional structures of phycobiliproteins. The figure generated from PyMOL software with PDB ID: 3O18 (*Thermosynechococcus vulcanus*)

of linker polypeptides to make a complete functional structure. Finally both rods and core cylinders associate together to form a complete PBS. The protein aggregates strongly, being found as a hexamer $(\alpha\beta)_6$ but also as a trimer $(\alpha\beta)_3$ and a monomer $(\alpha\beta)$ depending on the medium. Kobayashi et al. (1979) reported that conformation of PBPs could be optimized and maintained at monomer, trimer, and hexamer at pH 3.9, 5.6, and 8.0, respectively. Urea treatment on PC trimers $(\alpha\beta)_3$ resulted in monomerization $(\alpha\beta)$, which was followed by a complex unfolding process of the protein. PBPs are aggregated into two subunits (α and β) and the third subunit (γ), a linker peptide which is found in phycoerythrin. The gamma (γ) subunit is not universal and may exclusively be found in rod subunit of PE of certain marine cyanobacteria (*Synechococcus* sp.) and red algae (Ficner and Huber 1993; Six et al. 2005; Wilbanks and Glazer 1993). The structure of R-phycoerythrin (R-PE) can be described as $(\alpha\beta)_6\gamma$, while R-PC and R-APC have trimeric structure $(\alpha\beta)_3$ bound to specific cysteines by thioether bonds (Isailovic et al. 2004). Each PBP (Cyanophyta) is made up of heteromonomer of two subunits, α (12–20 kDa) and β (15–22 kDa) (Sinha et al. 1995a, b; Kannaujiya and Sinha 2016), whereas R-PE is made up of trimeric complex that includes α (18–20 kDa), β (19–21 kDa), and γ (30 kDa) (Galland-Irmouli et al. 2000).

R-PEs are commonly composed of 6α and 6β dimer subunits including 1γ subunit to make hexameric structure $(\alpha\beta)_3$-γ-$(\alpha\beta)_3$ with molecular mass up to 240–260 kDa (Chang et al. 1996; Rossano et al. 2003; Sun et al. 2004; Wang et al. 2015). Although, some red algae contain two or three γ subunits but their functional properties have not yet been ascertained (Su et al. 2010). In addition, some authors assume that γ subunits may play an important role in connecting trimers or hexamers of PE subunits (Wang et al. 2015). The hexameric form of R-PE isolated from *Polysiphonia urceolata* has 18–20 kDa molecular mass of α and β subunits and ~34 kDa of γ subunits (Chang et al. 1996).

Usually R-PEs exist in hexamer form in buffer solution; however, hexamer form of APC and PC are not stable in solution, and it generally exist in trimeric $(\alpha\beta)_3$ form (Jiang et al. 2001; Sun et al. 2003). Almost all red algae consisted of high amounts of PE among the total BPs; thus it plays an indispensable role in energy utilization and carbon storage in photosynthesis. The high content of PE ratio and

PE–PE similarity increases probability of coupling between multiple PE hexamers critical for assembly, disassembly, and stability of hexamer of rod domains (Wang et al. 2015). The protein aggregates strongly, being found as a hexamer $(\alpha\beta)_6$ but also as a trimer $(\alpha\beta)_3$ and a monomer $(\alpha\beta)$ depending on the medium. In the past, the entire structure of PBS has been studied at low resolution by electron microscopy (Gantt and Lipschultz 1972), but nowadays, X-ray crystallography has become a very good tool that provides high-resolution views of proteins. Crystal structures of PBS have also markedly enabled the details of the tertiary or hexameric structure of protein to identify the functionalities and stability under various environmental conditions. Several kinds of PBP structures have been resolved, such as PE (Ficner and Huber 1993; Jiang et al. 1999; Ritter et al. 1999), C-PC (Duerring et al. 1991), and APC (Brejc et al. 1995; Liu et al. 1999; Reuter et al. 1999). The crystal structures C-PC have been reported in many cyanobacteria; however, the crystal structure of R-PC is still unknown. R-PC-PU is the first crystal structure that has been reported in *Polysiphonia urceolata* which contain both phycoerythrobilin (PEB) and phycocyanobilin (PCB) (Jiang et al. 2001). A number of PBPs isolated from thermophile cyanobacteria that thrive at up to 65 °C are recently reported (Samsonoff and MacColl 2001). The research against thermal denaturation of PBPs has been well documented by changes in protein sequence as well as structural anomalies (Inoue et al. 2000; Fuglistaller et al. 1983; Sidler et al. 1981; Chen and Berns 1978, 1980).

3.3.3 Chromophores

PBPs have different numbers of chromophores, which are open-chain tetrapyrroles covalently bound to cysteines via thioether bonds. The chromophores can be classified by structure as phycoerythrobilin (PEB), phycocyanobilin (PCB), phycoviolobilin (PVB), or phycourobilin (PUB). Each biliprotein is made up of α and β subunit covalently attached to cysteines of the apoprotein via a thioether bond to C-3 on ring A and in some cases by an additional thioether bond to C-18 on ring D (Fig. 3.5). Cyanobacteria and red algae have diverse bilin chromophores such as PCB and PEB which possess a 3,3V-ethylidene group.

The chromophore synthesized from heme is attached to apoproteins of bilin by addition of a thiol group via ethylidene bond (Frankenberg et al. 2001; Wu and Lagarias 2000). Chromophores are generally bound to the polypeptide chain at conserved position either by one cysteinyl thioester linkage through the vinyl substituents on the pyrrole ring A of the tetrapyrrole or occasionally by two cysteine linkages through the vinyl substituents on both A and D pyrrole rings (Glazer 1985). The properties of light-harvesting PBPs in photosynthetic cyanobacteria are mainly dependent on chromophore assembly held in well-defined geometries inside the core of proteins (Zehetmayer et al. 2002). Dissociated subunits typically have less intense color and are mostly differentiated from native trimer to monomer subunits. In addition to generating biological activity, folding is coupled to many other biological processes, including the traffic of other molecules to specific cellular locations and the regulation of cellular growth and differentiation (Radford and Dobson

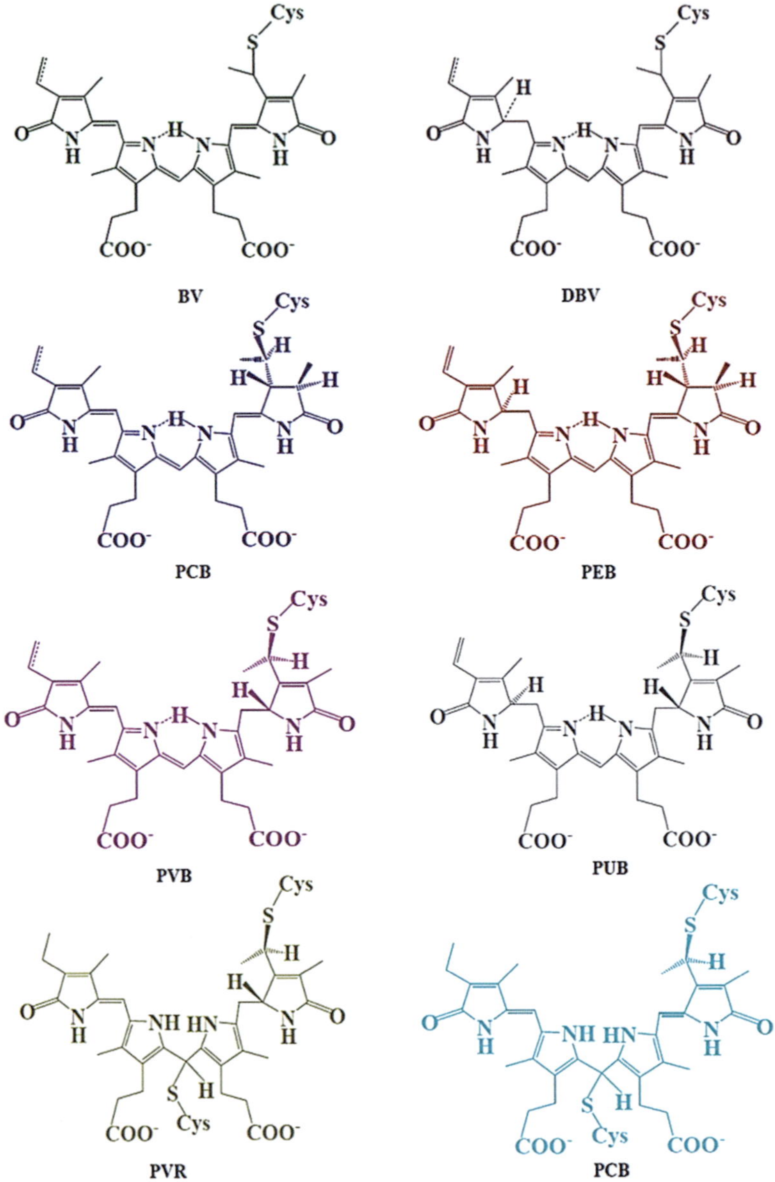

Fig. 3.5 Molecular sketch of cysteine-linked chromophores of phycobiliprotein complexes with their distinguished colors

1999). The electronic state of chromophores and associated apoproteins plays a determining role in the stable chemical structure of proteins (Kikuchi et al. 1997). Therefore, the chromophore acts to report the integrity of the protein structure; when in native form, the protein has a beautiful dark blue color, but upon denaturation the blue color fades away (MacColl and Berns 1981). Fluorescence techniques have been used for the investigation of PBPs, because it exhibits strong fluorescence when uncoupled from the reaction centers (Zehetmayer et al. 2002). When the tetrapyrrole chromophore is held in a linear conformation by the protein scaffolding, it has a fluorescence quantum yield of approximately 60 % with a fluorescence maximum of 650 nm in C-PC (Glazer 1984). C-phycocyanin (C-PC) carries three phycocyanobilin (PCB) chromophores in the heterodimeric (αβ) protomer, at cysteines α-84, β-84, and β-155. In addition, this residue is conserved in the β subunit categorized as Asp87 in β-84 and Asp39 present in β-155 chromophore of C-PC (Frank et al. 1978). However, it is also present in C-APC (Brejc et al. 1995) and C-PE (Sidler et al. 1986). The Asp87 residue is a key amino acid that determines the electronic state of the chromophore (Kikuchi et al. 1997). Scharnagl and Schneider (1989) suggested the protonation of chromophores resulting into red shifting in absorption maximum due to the presence of equal distance between Asp87 and the B- and C-rings of chromophores. Phycoerythrocyanin (PEC) is found in photosynthetic membrane of *Mastigocladus laminosus*. It is a complex mixture of PC and PE which comprises three heterodimeric substructures such as (αβ)-PEC (Duerring et al. 1991).The α subunit bears a single phycoviolobilin chromophore at position Cys84, whereas β subunit carries two phycocyanobilin chromophores at position Cys82 and Cys153. The α subunits (164 amino acid) and β subunits (177 amino acid) of C-PE, respectively, integrated with two to three covalently attached tetrapyrrole chromophores collectively called PEB (MacColl 1999). The computational analysis has reported two methodologies for detection and estimation of excitation energy transfer between chromophores. The detection of energy transfer primarily depends on two mechanisms such as strong coupling and another weak coupling between chromophores. In strong coupling, chromophores behave singly and only share delocalized energy, whereas in weak coupling, chromophores share electrons with each other and retain their emission spectra (MacColl 1998). It has been proved that the weak coupling between chromophores is a key mechanism for excitation energy transfer by the help of time-dependent density functional theory (Ren et al. 2006). Recently, Ren et al. (2013) have reported an additional pathway for excitation energy transfer from β^1-155 to α^2-84 (1497.8 ps) and α^2-84 to β^1-84 (0.4 ps) which is dramatically faster (up to 102–105 times) than other pathways in C-PC monomer and trimer.

On the basis of chromophore structure, R-PC can be divided into three groups: R-PC (I), R-PC (II), and R-PC (III) (Ong and Glazer 1988). All R-PCs carry three different kinds of chromophores at α84, β84, and β155 (R-PC (I) has α84PCB, β84PCB, and β155PEB; R-PC (II) has α84PEB, β 84PCB, and β155PEB; and R-PC (III) has α84PUB, β84PCB, and β155PCB). However, each subunit of R-PE is associated with five chromophores such as α84, β84, α140a, β155, and β50/61. However, R-PE subunits of *Gracilaria chilensis* have different α-PEB and β-PEB chromophores that are bound to C82, C139 and C82, C158 cysteine residues, respectively. In addition, one phycourobilin (PUB) is bound to C50–61 (βPUB50–61). APC carries only two chromophores located at α84 and β84 of the cysteine residue (Table 3.2).

Table 3.2 Chromophore binding sites in phycobiliproteins

	Chromophore binding sites on cysteine residue					
Types of biliproteins	α-75	α-84	α-140	β-50/60	β-84*	β-155
Allophycocyanin		PCB			PCB	
C-Phycocyanin		PCB			PCB	PCB
Phycoerythrocyanin		PVB			PCB	PCB
R-Phycoerythrin		PEB			PCB	PEB
R-Phycocyanin		PUB			PCB	PEB
Phycocyanin#		PUB			PCB	PCB
C-Phycoerythrin		PEB	PEB	PEB	PEB	PEB
CU-Phycoerythrin (PEI)*		PEB	PUB	PUB	PEB	PEB
CU-Phycoerythrin (PEI)*	PUB	PEB	PEB	PUB	PEB	PEB

For details, see the reference Tu (2008)
Consensus mark: #*Synechoccus* sp WH8102; **Synechoccus* sp WH8103

PBPs play a dominant role for capturing photonic energy by their continuous modulation of the ratio of rods (PE/PC) to adapt in various light qualities and quantities in changes in environmental condition. However, adaptations have become critical for PBPs when irradiation is very strong that causes heating effects and saturates the photosynthetic electron transport chain. Under these circumstances, excessive accumulation of overexcited chlorophyll molecules near the reaction centers leads to generation of harmful reactive oxygen species and damage to the photosynthetic system. It is believed that cyanobacteria have primary mitigation strategies to overcome the excessive excitation of PSII by non-photochemical quenching (NPQ), mediated by the orange carotenoid-binding protein (OCP), which effectively quenches PBP fluorescence and reduces the energy level in surroundings near the photosystem. The OCP-dependent blue light is independent of ΔpH and excitation pressure on thylakoid membrane. The detailed mechanism of OCP-dependent quenching has not been well understood (Kirilovsky 2007; Kirilovsky 2010).

3.3.4 Linker Polypeptides

The nonpigmented linker polypeptides are integrated into the PBP structure (Liu et al. 2005). The molecular masses of these polypeptides range from 8 to 120 kDa. These polypeptides help in stabilizing the PBP structure and determine the positions of specific sites in the complex. It also facilitates assembly of PBP-containing substructures with modulation in absorption characteristics of the PBPs to promote unidirectional transfer of energy from PBPs to physically link the entire complex to the Chl *a* (Bryant et al. 1991; Grossman et al. 1993). The classification of linker polypeptides is defined by specific abbreviations based on location and molecular mass (Glazer 1985). The common abbreviation of linker polypeptides is designated as L_X^Y where X refers to a position and Y refers to the molecular mass of the linker polypeptide (L) PBS complex. Moreover, on the basis of position, X can be designated as R (rod) and C (core) for main chain, while junction is represented as RC

Table 3.3 Types and characteristics of linker polypeptides

Protein	Symbol	Amino acids	MW (kDa)	pI	Annotation	Gene
CpeC	PE LR	285–294	31.8–33.1	9.6	PE-associated linker	*cpeC*
CpeD	PE LR	249–255	27.9–28.4	8.2–8.6	PE-associated linker	*cpeD*
CpeE	PE LR	244–254	27.1–28.4	9.7	PE-associated linker	*cpeE*
PecC	PEC LR	278–279	31.3–31.5	9.6–9.7	PEC-associated linker	*pecC*
CpcC	PC LR	219–291	24.8–32.6	9.5–9.6	PC-associated linker	*cpcC*
CpcD	PC LR	70–87	7.8–9.9	9.8–10.5	Rod capping linker	*cpcD*
CpcG	LRC	231–279	26.8–31.9	9.3–9.6	Rod–core linker	*cpcG*
CpcH	LRC	271–273	30.4–30.8	8.8–9.7	Rod–core linker	*cpcH*
CpcI	LRC	288	32.7	8.9	Rod–core linker	*cpcI*
ApcC	LC	66–69	7.7–7.8	10.9–11.4	APC-associated linker	*apcC*
ApcE	LCM	683–1155	76.5–129.8	9.5–9.7	Core–membrane linker	*apcE*

Adapted from Liu et al. (2005)

(rod–core) and CM (core–membrane). Most of the linker polypeptides are colorless, but at least two of them also carry covalently bound chromophores, namely, core–membrane linker (LCM) PCB-ApcE, and γ-subunits in class I and II of R-PE (Scheer and Zhao 2008). The details of PBP structure are quite variable in cyanobacteria and red algae. However, PBS is common in a central core that is composed of APC and specific linker polypeptides that are situated on the thylakoid membrane (photosystem II) of chlorophylls. Radiating out from the core is a number of rod elements composed of PCs (and often also PEs and other PBPs) together with their associated linker polypeptides.

The linker polypeptides are classified into four groups according to their position and functional properties PBS: Group I, L_R polypeptides (27–35 kDa) that participate in the assembly of the peripheral rods including few small rod linker polypeptides with 10 kDa (rod linkers such as LR^{10}, LR^{33}, and LR^{35} that associate trimeric or hexameric PC/PE substructures into rod segments); Group II, L_{RC} polypeptides (25–27 kDa) that participate in attaching the peripheral rods to the core subunits; Group III, LC polypeptides (8 kDa) that play a key role for attaching core components and their functional property; and Group IV, LCM (70–120 kDa) that is a larger molecular mass of polypeptides for attachment of PBS to photosynthetic membrane and acts as the major terminal energy emitter to PSII (Table 3.3) (Zilinskas and Greenwald 1986; de Lorimier et al. 1990; Capuano et al. 1991; Liu et al. 2005; Nganou et al. 2016).

The sequence analysis of linker polypeptides shows 75 % homology in the different types of PBPs. However, LC polypeptides show high sequence homology with the PBPs of different algae and indicate that they have more conserved sequences. The structural motif analysis of rod linker polypeptides is indicating six conserved domains (N-terminus) that play a significant role in packing and assembly of rod discs into hexamer (Lundell et al. 1981; Anderson and Toole 1998).The rod–core linker also possesses six conserved domains; the N-terminal occupies the central hole of rod–core and domain, while the C-terminal is comparatively less conserved which may help to connect distinct regions of core subunits. It is supposed that the N-terminal is buried in the central hole of trimers and protected with proteolytic treatments, whereas the C-terminal is buried in the hexamer for interconnection between rods and core (Parbel and Scheer 2000; Liu et al. 2005). LCM is the largest component of PBS, and it acts as an anchor polypeptide in core (APC) subunits. The structure of core subunits may change the molecular mass of LCM polypeptides such as bicylindrical, tricylindrical, and pentacylindrial cores (70–75 kDa, 92–99 kDa, and 115–128 kDa, respectively) (MacColl 2004; Zhao et al. 2005). The sequence analysis of LCM has defined several domains including N-terminal PBPs and L_RPC domains that may support for hexamer formation in APC core (Capuano et al. 1991). The C-terminal domain of the LCM includes REP domains (repeat domains) which are involved in the interaction within APC and assembly of the PBS core (Bryant et al. 1991). The surfaces of linker polypeptides are positively charged, whereas most of the globular proteins are likely to be hydrophobic; thus both interactions occur between PBPs and linker polypeptides (Wilk et al. 1999).

3.4 Chromophore Integrity and Energy Pathway

They are associated with variable electronic coupling in chromophores; thereby they have wide absorption property and efficient transfer of energy (Zehetmayer et al. 2002). The electrostatic interaction between basic polypeptides and acidic polypeptides significantly stabilizes the PBP assembly and facilitates the unidirectional transport of energy toward the photosynthetic reaction center (Tandeau de Marsac and Cohen-Bazire 1977). All the subunits have unique amino acid composition to make a complete functional structure. Aromatic amino acid such as tyrosine is dominated in α and β subunits of PBPs. Tyrosine intrinsic anisotropy and fluorescence lifetime is optimal to characterize nanosecond and sub-nanosecond motions in peptides and proteins (Ferreira et al. 1994), and therefore, it could be a useful tool to study structural and dynamic changes in peptides upon interaction with membranes. Redox state is known to regulate the structural and functional integrity of PBPs. Cysteine residues have unique sulfhydryl groups that maintain the stability of chromophores in various stresses. Unfortunately, to the best of our knowledge, no work has been done on the sulfhydryl groups of the cysteine molecule. Proper redox state of sulfhydryl groups is required to maintain structural integrity of chromophores inside the PBPs. Therefore, an arrangement of chromophores within the PBPs allows absorption and unidirectional transfer of the light energy to the chlorophyll of PSII pigment proteins in thylakoid membrane. Efficient energy transfer from the PBPs is restricted up to 95% due to their geometrical arrangement. Some evidences show that PBS is directly associated with membrane pigment proteins possibly from PSII (Bryant et al. 1979).

3.4.1 Energy Transfer Mechanism in Phycobilisomes

The molecular analysis of PBS crystal structure is a promising method for the determination of internal excitation energy transfer process between chromophore and its associated subunits. Till date, several crystal structures of diverse composition of PBPs have been analyzed, and the stoichiometric arrangement of chromophores for possible transfer of energy has been located (Wehrmeyer 1983; Bryant et al. 1990; Glauser et al. 1992; Wang et al. 2001; Jiang et al. 2001; Doust et al. 2004; Contreras-Martel et al. 2007; David et al. 2011; Camara-Artigas et al. 2012; Marx and Adir 2013; David et al. 2014; Fromme et al. 2015). The mechanism of transfer of energy is still unclear due to uneven distance of chromophores in different PBPs (Fig. 3.6).

Trimeric PBPs are disc-like structures with a face which may participate in hexamer formation, although the backside of disc is not participating in hexamer formation. Thus, front portion of disc play a keen role in the formation of hexameric functional structure. The electron microscopy observation reveals different types of assembly in PBPs (Yamanaka et al. 1980). There are five types of assembly that

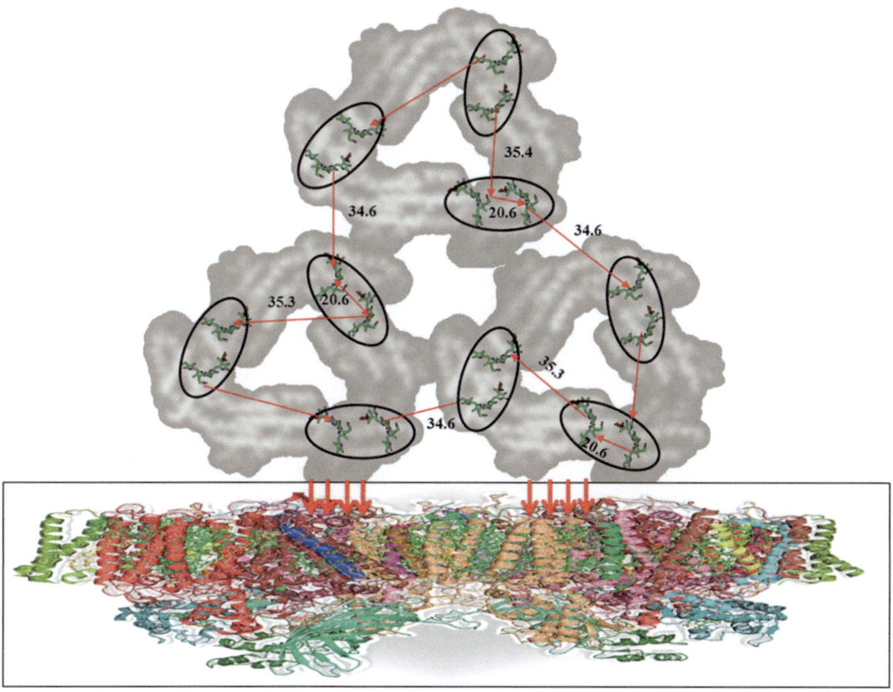

Fig. 3.6 Schematic representation of excitation energy transfer in the three-dimensional cylindrical core of PBSs with one of the four disc trimers stacked in each cylinder, one upper and two lower. Chromophores (six in each disc trimer) are shown in black against the background of apoproteins (For details, readers are suggested to see the references Loll et al. 2005; Stadnichuk et al. 2015b; Kannaujiya et al. 2016)

occurs between PE, PC, and APC subunits of PBPs (Jiang et al. 2001). The first type of assembly is the back-to-back association between PE and PC, whereas the second type of assembly is parallel side-to-side assembly. The third type of assembly is generated by the back-to-side association between PC and APC, whereas the fourth type of assembly is associated with side-to-side assembly in perpendicular manner. The fifth assembly is observed in PC to PC by side-to-side perpendicular to type II assembly.

3.4.1.1 Energy Transfer PE to PC

Structurally, back-to-back and side-to-side assembly occur between PE and PC. The back-to-back assembly is the main assembly of hexamer of R-PE and R-PC which consisted of six pairs of chromophore–chromophore distances shorter than 40 Å (Jiang et al. 2001). The six pairs of chromophores such as PE-β84 to PC-α84 (35 Å), PE-β84 to PC-β84 (27 Å), PE-α84 to PC-β84 (35 Å), PE-α84 to PC-α84 (32 Å), PE-β50 to PC-α84 (27 Å), and PE-β50 to PC-β84 (35 Å) are used for transfer of efficient energy between chromophores (Jiang et al. 2001; Kannaujiya et al. 2016). However, some assemblies such as PEB-β155 and PEB-α140a (PE) and β155 (PC) are not incorporated in back-to-back assemblies; otherwise it may participate parallel side-to-side assembly for energy transfer (Jiang et al. 1999). The 17-ps or 20-ps component was assigned to PC-α84 or PC-β84 to APC, whereas the 55-ps component was assigned to PC-β155 to APC by the mediation of PC-α84 or PC-β84. The side-to-side assemblies have quite large distance (>40 Å) between the chromophores of trimers; thus it may not efficiently transfer the energy between chromophores.

3.4.1.2 Energy Transfer PC to APC

The main assembly for way of energy transfer between PC (back) and APC (side) is well established (Debreczeny and Sauer 1995; Sandstrom et al. 1988). The cyanobacterium *Acaryochloris marina* has developed two different types of accessory light-harvesting complexes containing chlorophyll-d and common PBPs present in almost every cyanobacteria. The conformation of PBPs is a heterohexamer made up of trimeric subunits of PC and APC (Theiss et al. 2011). The excitation energy transfer from PC to Chl d is generally characterized by four kinetic components with lifetimes of <400 fs, 3 ps, 14 ps, and 70 ps at room temperature (Gryliuk et al. 2014).

3.4.2 Energy Transfer Mechanism in Photosystem

3.4.2.1 Energy Transfer APC to PS

The growth of oxygen-evolving organisms such as cyanobacteria/algae is keenly dependent on excitation energy transfer for carbon storage in photosynthetic machinery. Arba et al. (2013) have reported that the growth of *Arthrospira platensis* is dependent on excitation energy transfer between PC, APC, and PSII including PSII red pigmented chlorophyll (F685 and F695) and PSI including PSI red pigmented chlorophyll (F730 and F760) (Boussiba and Richmond 1979; Shubin et al. 1991). The red chlorophylls exist in the photosynthetic antenna complex attached to the reaction center and efficiently harvest long-wavelength radiation *via* direct energy transfer from PBPs to PSI or indirect transfer from PBPs to PSII to PSI (Gobets and van Grondelle 2001; Akimoto et al. 2012). The changes in growth medium of *Arthrospira platensis* have also changed the pathway of excitation energy transfer (Arba et al. 2013). The complex dynamics of excitation energy transfer in the PBPs of *A. marina* is revealed to be up to femtosecond (fs) in resolution (Theiss et al. 2008). The detailed understanding of excitation energy transfer in PBP occurs by time-resolved spectroscopy for molecular investigations of states of excited electronic and electro-vibration coupling after excitation at specific wavelength. Generally, pigment–protein complexes are amorphous in nature; thus detail of spectrum could not understand due to homogeneous broadening of peaks (Jankowiak et al. 2011). As recently reported, some advanced techniques such as spectral hole burning (SHB) (Jankowiak et al. 2011), difference fluorescence line-narrowing (ΔFLN), and fluorescence line-narrowing (FLN) efficiently inhibited the inhomogeneous broadening of peaks (Fünfschilling et al. 1986; Rätsep and Freiberg 2007). The information of electron–vibrational coupling can be gathered from non-resonant vibrational satellite holes (Gillie et al. 1989; Gryliuk et al. 2014). SHB technique has been already applied to investigate the different subunits of cyanobacterial PBPs such as PE, PC, and APC (Friedrich et al. 1981a, b).

3.5 Conclusion

The diversity of ancient accessory light-harvesting complex among Cyanophyta, Rhodophyta, and Cryptophyta always facilitates the research and development to understand the hidden functional aspect of energy utilization. Structural significance of PBS including PBPs, LPs, and associated chromophores is playing a primary role of harvesting light energy from the sun. The immense significance of PBS includes over 1500 articles, monographs, and several crystal structures (RCSB-PDB database) which have been recommended as a primary source to understand the structural property. The continuous progress in genome sequencing further combined PBS analysis at genomic level. Unfortunately, in present scenario, our understanding about linker polypeptides, chromophores, and excitation energy transfer from PBS to PS is still concealed in diverse species. In the present study, we have made an attempt to study in detail about structure and function of PBS including PBPs and LPs in diverse micro-/macroalgal species. The recent advances on chromophore research and its role in excitation energy transfer have been discussed. The importance of linker polypeptides for the assembly of defined complexes and their role in stability of PBPs has been discussed. Future research will drive the field toward a greater understanding of the evolutional mechanisms of PBPs and LPs in cyanobacteria and red algae that can be put to proper utilization in various applications in biological science including energy devices.

References

Adir N (2005) Elucidation of the molecular structures of components of the phycobilisome: reconstructing a giant. Photosynth Res 85:15–32

Adir N, Dines M, Klartag M, McGregor A, Melamed-Frank M (2006) Assembly and disassembly of phycobilisomes. In: Shively JM (ed) Microbiology monographs: inclusions in prokaryotes. Springer, Berlin, pp 47–77

Akimoto S, Yokono M, Hamada F, Teshigahara A, Aikawa S, Kondo A (2012) Adaptation of light-harvesting systems of *Arthrospira platensis* to light conditions, probed by time-resolved fluorescence spectroscopy. Biochim Biophys Acta 1817:1483–1490

Allen JF, de Paula WBM, Puthiyaveetil S, Nield J (2011) A structural phylogenetic map for chloroplast photosynthesis. Trends Plant Sci 16:645–655

Ana AA, Ghada A, Egbert JB (2009) Structural organization of phycobilisomes from *Synechocystis* sp. strain PCC6803 and their interaction with the membrane. Biochim Biophys Acta 1787:272–279

Anderson LK, Toole CM (1998) A model for early events in the assembly pathway of cyanobacterial phycobilisomes. Mol Microbiol 30:467–474

Apt KE, Collier JL, Grossman AR (1995) Evolution of the phycobiliproteins. J Mol Biol 248:79–96

Arba M, Aikawa S, Niki K, Yokono M, Kondo A, Akimoto S (2013) Differences in excitation energy transfer of *Arthrospira platensis* cells grown in seawater medium and freshwater medium, probed by time-resolved fluorescence spectroscopy. Chem Phys Lett 588:231–236

Arteni AA, Ajlani G, Boekema EJ (2009) Structural organization of phycobilisomes from *Synechocystis* sp. strain PCC6803 and their interaction with the membrane. Biochim Biophys Acta 1787:272–279

Bermejo R, Acien FG, Ibanez MJ, Fernandez JM, Molina E, Alvarez-Pez JM (2003) Preparative purification of B-phycoerythrin from the microalga *Porphyridium cruentum* by expanded-bed adsorption chromatography. J Chromatogr B 790:317–325

Boussiba S, Richmond AE (1979) Isolation and purification of phycocyanins from the blue green algae *Spirulina platensis*. Arch Microbiol 120:155–159

Brejc K, Ficner R, Huber R, Steinbacher S (1995) Isolation, crystallization, crystal structure analysis and refinement of allophycocyanin from cyanobacterium *Spirulina platensis* at 2.3 Å resolution. J Mol Biol 249:424–440

Bryant DA, Guglielmi G, Tandeau de Marsac N, Castlets AM, Cohen-Bazire G (1979) The structure of cyanobacterial phycobilisomes: a model. Arch Microbiol 123:113–127

Bryant DA, De Loremier R, Guglielme G, Stevens SE (1990) Structural and compositional analyses of the phycobilisomes of *Synechococcus* sp. PCC 7002: analyses of the wild-type strain and a phycocyanin- less mutant constructed by interposon mutagenesis. Arch Microbiol 153:550–560

Bryant DA, Stirewalt VL, Glauser M, Frank G, Sidler W, Zuber H (1991) A small multigene family encodes the rod-core linker polypeptides of *Anabaena* sp. PCC 7120 phycobilisomes. Gene 107:91–99

Camara-Artigas A, Bacarizo J, Andujar-Sanchez M, Ortiz-Salmeron E, Mesa-Valle C, Cuadri C, Martin-Garcia JM, Martinez-Rodriguez S, Mazzuca-Sobczuk T, Ibanez MJ, Allen JP (2012) pH-dependent structural conformations of B-phycoerythrin from *Porphyridium cruentum*. FEBS J 279:3680–3691

Capuano V, Braux AS, Tandeau de Marsac N, Houmard J (1991) The anchor polypeptide of cyanobacterial phycobilisomes. Molecular characterization of the *Synechococcus* sp. PCC 6301 *apcE* gene. J Biol Chem 266:7239–7247

Chang W, Jiang T, Wan Z, Zhang J, Yang Z, Liang D (1996) Crystal structure of R-phycoerythrin from *Polysiphonia urceolata* at 2.8 Å resolution. J Mol Biol 262:721–731

Chen CH, Berns DS (1978) Comparison of the stability of phycocyanins from thermophilic, mesophilic, psychrophilic and halophilic algae. Biophys Chem 8:203–213

Chen CH, Berns DS (1980) Thermotropic properties of thermophilic, mesophilic, and psychrophilic bluegreen algae. Plant Physiol 66:596–599

Chen M, Li Y, Birch D, Willows RD (2012) A cyanobacterium that contains chlorophyll f-a red-absorbing photopigment. FEBS Lett 586:3249–3254

Contreras-Martel C, Matamala A, Bruna C, Poo-Caama G, Almonacid D, Figueroa M, Martinez-Oyanedel J, Bunster M (2007) The structure at 2 Å resolution of phycocyanin from *Gracilaria Chilensis* and the energy transfer network in a Pc-Pc complex. Biophys Chem 125:388

David L, Marx A, Adir N (2011) High-resolution crystal structures of trimeric and rod phycocyanin. J Mol Biol 405:201–213

David L, Prado M, Arteni AA, Elmlund DA, Blankenship RE, Adir N (2014) Structural studies show energy transfer within stabilized phycobilisomes independent of the mode of rod-core assembly. Biochim Biophys Acta 1837:385–395

de Lorimier R, Bryant DA, Stevens SE (1990) Genetic analysis of a 9 kDa phycocyanin-associated linker polypeptide. Biochim Biophys Acta 1019:29–41

Debreczeny MP, Sauer K (1995) Comparison of calculated and experimentally resolved rate constants for excitation energy transfer in C-phycocyanin. II. Trimers. J Phys Chem 99:8420–8431

Doust AB, Marai CNJ, Harrop SJ, Wilk KE, Curmi PMG, Scholes GD (2004) Developing a structure-function model for the cryptophyte phycoerythrin 545 using ultrahigh resolution crystallography and ultrafast laser spectroscopy. J Mol Biol 344:135–153

Doust AB, Wilk KE, Curmi PMG, Scholes GD (2006) The photophysics of cryptophyte light-harvesting. J Photochem Photobiol A Chem 184:1–17

Duerring M, Schmidt GB, Huber R (1991) Isolation, crystallization, crystal structure analysis and refinement of constitutive C-phycocyanin from the chromatically adapting cyanobacterium *Fremyella diphosiphon* at 1.66 Å resolution. J Mol Biol 217:577–592

Ferreira ST, Stella L, Gratton E (1994) Conformational dynamics of bovine Cu, Zn superoxide dismutase revealed by time-resolved fluorescence spectroscopy of the single tyrosine residue. Biophys J 66:1185–1196

Ficner R, Huber R (1993) Refined crystal structure of phycoerythrin from *Porphyridium cruentum* at 0.23 nm resolution and localization of the gamma-subunit. Eur J Biochem 218:103–106

Fischer WF (2008) Biogeochemistry: life before the rise of oxygen. Nature 455:1051–1052

Frank G, Sidler WA, Widmer H, Zuber H (1978) The complete amino acid sequence of both subunits of C-phycocyanin from the cyanobacterium *Mastigocladus laminosus*. Hoppe Seylers Z Physiol Chem 359:1491–1507

Frankenberg N, Mukougawa K, Kohchi T, Lagarias JC (2001) Functional genomic analysis of the hy2 family of ferredoxin-dependent bilin reductases from oxygenic photosynthetic organisms. Plant Cell 13:965–978

Friedrich J, Scheer H, Zickendraht-Wendelstadt B, Haarer D (1981a) Photochemical hole burning: a means to observe high resolution optical structures in phycoerythrin. J Chem Phys 74:2260–2266

Friedrich J, Scheer H, Zickendraht-Wendelstadt B, Haarer D (1981b) High-resolution optical studies on C-phycocyanin via photochemical hole burning. J Am Chem Soc 103:1030–1035

Fromme R, Ishchenko A, Metz M, Roy-Chowdhury S, Basu S, Boutet S, Fromme P, White TA, Barty A, Spence JCH, Weierstall U, Liu W, Cherezov V (2015) Serial femtosecond crytallography of soluble proteins in lipidic cubic phase. IUCrJ 2:545

Fuglistaller P, Suter F, Zuber H (1983) The complete amino-acid sequence of both subunits of phycoerythrocyanin from the thermophilic cyanobacterium *Mastigocladus laminosus*. Hoppe Seylers Z Physiol Chem 364:691–712

Fünfschilling J, Glatz D, Zschokke-Gränacher I (1986) Hole-burning spectroscopy as a tool to eliminate inhomogeneous broadening. J Lumin 36:85–92

Galland-Irmouli AV, Pons L, Lucon M, Villaume C, Mrabet NT, Gueant JL, Fleurence J (2000) One-step purification of R-phycoerythrin from the red macroalga *Palmaria palmata* using preparative polyacrylamide gel electrophoresis. J Chromatogr B Biomed Sci Appl 739:117–123

Gantt E, Lipschultz CA (1972) Phycobilisomes of *Porphyridium cruentum*: I. Isolation. J Cell Biol 54:313–324

Gantt E, Grabowski B, Cunningham FX Jr (2003) Antenna systems of red algae: phycobilisomes with photosystem II and chlorophyll complexes with photosystem I. In: Green BR, Parson WW (eds) Light harvesting antennas in photosynthesis. Kluwer Academic Publishers, Dordrecht, pp 307–322

Gillie JK, Small GJ, Golbeck JH (1989) Non photochemical hole burning of the native antenna complex of photosystem I (PSI-200). J Phys Chem 93:1620–1627

Glauser M, Bryant DA, Frank G, Wehrli E, Rusconi SS, Sidler W, Zuber H (1992) Phycobilisome structure in the cyanobacteria *Mastigocladus laminosus* and *Anabaena* sp. *PCC* 7120. Eur J Biochem 205:907–915

Glazer AN (1984) Phycobilisome: a macromolecular complex optimized for light energy transfer. Biochim Biophys Acta 768:29–51

Glazer AN (1985) Light harvesting by phycobilisome. Annu Rev Biophys Chem 14:47–77

Glazer AN (1989) Light guides. J Biol Chem 264:1–4

Gobets B, van Grondelle R (2001) Energy transfer and trapping in photosystem I. Biochim Biophys Acta 1507:80–99

Grossman A, Schaefer MR, Chiang GG, Collier JL (1993) The phycobilisomes, a light-harvesting complex responsive to environmental conditions. Microbiol Rev 57:725–749

Gryliuk G, Rätsep M, Hildebrandt S, Irrgang K-D, Eckert H-J, Pieper J (2014) Excitation energy transfer and electron-vibrational coupling in phycobiliproteins of the cyanobacterium *Acaryochloris marina* investigated by site-selective spectroscopy. Biochim Biophys Acta 1837:1490–1499

Guglielmi G, Cohen-Bazire G, Bryant DA (1981) The structure of *Gloeobacter violaceus* and its phycobilisomes. Arch Microbiol 129:181–189

Inoue N, Emi T, Yamane Y, Kashino Y, Koike H, Satoh K (2000) Effects of high-temperature treatments on a thermophilic cyanobacterium *Synechococcus vulcanus*. Plant Cell Physiol 41:515–522

Isailovic D, Li HW, Yeung ES (2004) Isolation and characterization of R-phycoerythrin subunits and enzymatic digests. J Chromatogr A 1051:119–130

Jankowiak R, Reppert M, Zazubovich V, Pieper J, Reinot T (2011) Site selective and single complex laser-based spectroscopies: a window on excited state electronic structure, excitation energy transfer, and electron-phonon coupling of selected photosynthetic complexes. Chem Rev 8:4546–4598

Jiang T, Zhang JP, Liang DC (1999) Structure and function of chromophores in R-phycoerythrin at 1.9 Å resolution. Proteins Struct Funct Genet 34:224–231

Jiang T, Zhang J, Chang W, Liang D (2001) Crystal structure of R-phycocyanin and possible energy transfer pathways in the phycobilisome. Biophys J 81:1171–1179

Johnson EM, Kumar K, Das D (2014) Physicochemical parameters optimization, and purification of phycobiliproteins from the isolated *Nostoc* sp. Bioresour Technol 166:541–547

Kannaujiya VK, Sinha RP (2016) An efficient method for the separation and purification of phycobiliproteins from a rice-field cyanobacterium *Nostoc* sp. strain HKAR-11. Chromatographia 79:335–343

Kannaujiya VK, Rahman A, Adinath, Sundaram S, Sinha RP (2016) Structural and functional dynamics of tyrosine amino acid in phycocyanin of hot-spring cyanobacteria: a possible pathway for internal energy transfer. Gene Rep 5:83–91

Kikuchi H, Sugimoto T, Mimuro M (1997) An electronic state of the chromophore, phycocyanobilin, and its interaction with the protein moiety in C-phycocyanin: protonation of the chromophore. Chem Phys Lett 274:460–465

Kirilovsky D (2007) Photoprotection in cyanobacteria: the orange carotenoid protein (OCP)-related non-photochemical-quenching mechanism. Photosynth Res 93:7–16

Kirilovsky D (2010) The photoactive orange carotenoid protein and photoprotection in cyanobacteria. Adv Exp Med Biol 675:139–159

Kobayashi T, Degenkolb O, Bersohn R, Retzepis PM, MacColl R, Bern DS (1979) Energy transfer among the chromophores in phycocyanins measured by picoseconds kinetics. Biochemistry 18:5073–5078

Koyama K, Tsuchiya T, Akimoto S, Yokono M, Miyashita H, Mimuro M (2006) New linker proteins in phycobilisomes isolated from the cyanobacterium *Gloeobacter violaceus* PCC 7421. FEBS Lett 580:3457–3461

Li Y, Lin Y, Garvey CJ, Birch D, Corkery RW, Loughlina PC, Scheer H, Willows RD, Chena M (2016) Characterization of red-shifted phycobilisomes isolated from the chlorophyll f-containing cyanobacterium *Halomicronema hongdechloris*. Biochim Biophys Acta 1857:107–114

Liu JY, Jiang T, Zhang JP, Liang DC (1999) Crystal structure of allophycocyanin from red algae *Porphyra yezoensis* at 2.2 Å resolution. J Biol Chem 274:16945–16952

Liu L-N, Chen X-L, Zhang Y-Z, Zhou B-C (2005) Characterization, structure and function of linker polypeptides in phycobilisomes of cyanobacteria and red algae: an overview. Biochim Biophys Acta 1708:133–142

Loll B, Kern J, Saenger W, Zouni A, Biesiadka J (2005) Towards complete cofactor arrangement in the 3.0 A resolution structure of photosystem II. Nature 438:1040–1044

Lundell DJ, Williams RC, Glazer AN (1981) Molecular architecture of a light-harvesting antenna. In vitro assembly of the rod substructures of *Synechococcus* 6301 phycobilisomes. J Biol Chem 256:3580–3592

MacColl R (1998) Cyanobacterial phycobilisomes. Struct Biol 1998:124–311

MacColl R (1999) Studies on R-phycoerythrins from two Antarctic marine red algae and a mesophilic red alga. Polar Biol 22:384–388

MacColl R (2004) Allophycocyanin and energy transfer. Biochim Biophys Acta 1657:73–81

MacColl R, Berns D (1981) Phycocyanin in physical-chemical studies spectra and excitation-energy transfer. Isr J Chem 21:296–300

Marquardt J, Senger H, Miyashita H, Miyachi S, Morschel E (1997) Isolation and characterization of biliprotein aggregates from *Acaryochloris marina*, a Prochloron-like prokaryote containing mainly chlorophyll d. FEBS Lett 410:428–432

Marx A, Adir N (2013) Allophycocyanin and phycocyanin crystal structures reveal facets of phycobilisome assembly. Biochim Biophys Acta 1827:311–318

McGregor A, Klartag M, David L, Adir N (2008) Allophycocyanin trimer stability and functionality are primarily due to polar enhanced hydrophobicity of the phycocyanobilin binding pocket. J Mol Biol 384:406–421

Nganou C, David L, Adir N, Mkandawire M (2016) Linker proteins enable ultrafast excitation energy transfer in the phycobilisome antenna system of *Thermosynechococcus vulcanus*. Photochem Photobiol Sci 15:31–44

Ong LJ, Glazer AN (1988) Structural studies of phycobiliproteins in unicellular marine cyanobacteria. In: Stevens SE Jr, Bryant DA (eds) Light-energy transduction in photosynthesis: higher plant and bacterial models. The American Society of Plant Physiologists, Rockville, pp 102–121

Ort DR et al (2015) Redesigning photosynthesis to sustainably meet global food and bioenergy demand. Proc Natl Acad Sci U S A 112(28):8529–8536

Parbel A, Scheer H (2000) Model for the phycobilisome rod with interlocking disks based on domain-weighted linker-polypeptide sequence homologies of *Mastigocladus laminosus*. Int J Photoenergy 2:31–40

Price DC, Chan CX, Yoon HS, Yang EC, Qiu H, Bhattacharya D (2012) *Cyanophora paradoxa* genome elucidates origin of photosynthesis in algae and plants. Science 335:843–847

Radford SE, Dobson CM (1999) From computer simulations to human disease: emerging themes in protein folding. Cell 97:291–298

Rastogi RP, Sinha RP (2009) Biotechnological and industrial significance of cyanobacterial secondary metabolites. Biotechnol Adv 27:521–539

Rastogi RP, Kumari S, Richa HT, Sinha RP (2012) Molecular characterization of hot spring cyanobacteria and evaluation of their photoprotective compounds. Can J Microbiol 58:719–727

Rätsep M, Freiberg A (2007) Electron-phonon and vibronic couplings in the FMO bacteriochlorophyll *a* antenna complex studied by difference fluorescence line narrowing. J Lumin 127:251–256

Ren YL, Wan J, Xu X, Zhang QY, Yang GF (2006) A time-dependent density functional theory investigation of the spectroscopic properties of the beta-subunit in C-phycocyanin. J Phys Chem B 110:18665–18669

Ren Y, Chi B, Melhem O, Wei K, Feng L, Li Y, Han X, Li D, Zhang Y, Wan J, Xu X, Yang M (2013) Understanding the electronic energy transfer pathways in the trimeric and hexameric aggregation state of cyanobacteria phycocyanin within the framework of Förster theory. J Comput Chem 34:1005–1012

Reuter W, Wiegand G, Huber R, Than ME (1999) Structural analysis at 2.2 Å of orthorhombic crystals presents the asymmetry of the allophycocyanin-linker complex, AP.LC7.8, from phycobilisomes of *Mastigocladus laminosus*. Proc Natl Acad Sci U S A 96:1363–1368

Ritter S, Hiller RG, Wrench PM, Welte W, Diederichs K (1999) Crystal structure of a phycourobilin-containing phycoerythrin at 1.90 Å resolution. J Struct Biol 126:86–97

Rodríguez H, Rivas J, Guerrero MG, Manuel LM (1991) Enhancement of phycobiliprotein production in nitrogen-fixing cyanobacteria. J Biotechnol 20:263–270

Rosinski J, Hainfeld JF, Rigbi M, Siegelman HW (1981) Phycobilisomes ultrastructure and chromatic adaptation in *Fremyella diplosihon*. Ann Bot 47:1–12

Rossano R, Ungaro N, Ambrosio AD, Liuzzi GM, Riccio P (2003) Extracting and purifying R-phycoerythrin from Medieterranean red algae *Corallina elongata* Ellis & Solander. J Biotechnol 101:289–293

Samsonoff WA, MacColl R (2001) Biliproteins and phycobilisomes from cyanobacteria and red algae at the extremes of habitat. Arch Microbiol 176:400–405

Sandstrom A, Gillbro T, Sundstrom V, Xendler J, Holzwarth AR (1988) Picosecond study of energy transfer within 18-S particles of AN 112 (a mutant of *Synechococcus* 6301) phycobilisomes. Biochim Biophys Acta 933:54–64

Santiago-Santos MC, Ponce-Noyola T, Olvera-Ramírez R, Ortega-López J, Cañizares-Villanueva RO (2004) Extraction and purification of phycocyanin from *Calothrix* sp. Process Biochem 39:2047–2052

Scharnagl C, Schneider S (1989) UV-visible absorption and circular dichroism spectra of the subunits of C-phycocyanin. I: quantitative assessment of the effect of chromophore-protein interaction in the α-subunit. J Photochem Photobiol B Biol 3:603–615

Scheer H, Zhao KH (2008) Biliprotein maturation: the chromophore attachment. Mol Microbiol 68:263–276

Schirmer T, Bode W, Huber R, Sidler W, Zuber H (1985) X-ray crystallographic structure of the light-harvesting biliprotein C-phycocyanin from the thermophilic cyanobacterium *Mastigocladus laminosus* and its resemblance to globin structures. J Mol Biol 184:257–277

Shubin VV, Murthy SDS, Karapetyan NV, Mohanty P (1991) Origin of the 77 K variable fluorescence at 758 nm in the cyanobacterium *Spirulina platensis*. Biochim Biophys Acta 1060:28–36

Sidler WA (1994) Phycobilisome and phycobiliprotein structures. In: Bryant DA (ed) The molecular biology of cyanobacteria. Kluwer Academic Publication, The Netherlands, pp 139–216

Sidler W, Gysi J, Isker E, Zuber H (1981) The complete amino acid sequence of both subunits of allophycocyanin, a light harvesting protein–pigment complex from the cyanobacterium *Mastigocladus laminosus*. Hoppe Seylers Z Physiol Chem 362:611–628

Sidler W, Kumpf B, Rüdiger W, Zuber H (1986) The complete amino-acid sequence of C-phycoerythrin from the cyanobacterium *Fremyella diplosiphon*. Hoppe Seyler Z Biol Chem 367:627–642

Singh SP, Montgomery BL (2013) Salinity impacts photosynthetic pigmentation and cellular morphology changes by distinct mechanisms in *Fremyella diplosiphon*. Biochem Biophys Res Commun 433:84–89

Singh NK, Sonani RR, Rastogi RP, Madamwar D (2015) The phycobilisomes: an early requisite for efficient photosynthesis in cyanobacteria. EXCLI J 14:26–289

Sinha RP, Häder D-P (1996) Photobiology and ecophysiology of rice field cyanobacteria. Photochem Photobiol 64:887–896

Sinha RP, Lebert M, Kumar A, Kumar HD, Häder D-P (1995a) Spectroscopic and biochemical analyses of UV effects of phycobilisomes of *Anabaena* sp. and *Nostoc carmium*. Bot Acta 108:87–92

Sinha RP, Lebert M, Kumar A, Kumar HD, Häder D-P (1995b) Disintegration of phycobilisomes in a rice field cyanobacterium *Nostoc* sp. following UV irradiation. Biochem Mol Biol Int 37:697–706

Six C, Thomas JC, Thion L, Lemoine Y, Zal F, Partensky F (2005) Two novel phycoerythrin-associated linker proteins in the marine cyanobacterium *Synechococcus* sp. strain WH8102. J Bacteriol 187:1685–1694

Six C, Thomas JC, Garczarek L, Ostrowski M, Dufresne A, Blot N et al (2007) Diversity and evolution of phycobilisomes in marine *Synechococcus* spp.: a comparative genomics study. Genome Biol 2007:R259

Spear-Bernstein L, Miller KR (1989) Unique location of the phycobiliprotein light-harvesting pigment in the Cryptophyceae. J Phycol 25:412–419

Stadnichuk IN, Krasilnikov PM, Zlenko DV (2015a) Cyanobacterial phycobilisomes and phycobiliproteins. Microbiology 84:101–111

Stadnichuk VI, Lukashev EP, Yanyushin MF, Zlenko DV, Muronez EM, Stadnichuk IN, Krasilnikov PM (2015b) Energy transfer pathways among phycobilin chromophores and fluorescence emission spectra of the phycobilisome core at 293 and 77 K. Dokl Biochem Biophys 465:401–405

Stanier RY, Cohen-Bazire G (1977) Phototrophic prokaryotes: the cyanobacteria. Annu Rev Microbiol 31:225–274

Su H-N, Xie B-B, Zhang X-Y, Zhou B-C, Zhang Y-Z (2010) The supramolecular architecture, function, and regulation of thylakoid membranes in red algae: an overview. Photosynth Res 106:73–87

Sun L, Wang S (2011) Phycobiliprotein components and characteristics of the phycobilisome from a thermophilic cyanobacterium *Myxosarcina concinna* Printz. J Phycol 47:1304–1315

Sun L, Wang S, Chen L, Gong X (2003) Promising fluorescent probes from phycobiliproteins. IEEE J Sel Top Quantum Electron 9:177–188

Sun L, Wang S, Gong X, Chen L (2004) A rod-linker-contained R-phycoerythrin complex from the intact phycobilisome of the marine red alga *Polysiphonia urceolate*. J Photochem Photobiol B Biol 76:1–11

Sun L, Wang S, Gong X, Zhao M, Fu X, Wanga L (2009) Isolation, purification and characteristics of R-phycoerythrin from a marine macroalga *Heterosiphonia japonica*. Protein Expr Purif 64:146–154

Tandeau de Marsac N, Cohen-Bazire G (1977) Molecular composition of cyanobacterial phycobilisomes. Proc Natl Acad Sci U S A 74:1635–1639

Theiss C, Schmitt FJ, Andree S, Cardenas-Chavez C, Wache K, Fuesers J, Vitali M, Wess M, Kussin S, Eichler HJ, Eckert H-J (2008) Excitation energy transfer in the phycobiliprotein antenna of *Acaryochloris marina* studied by transient fs absorption and fluorescence spectroscopy, photosynthesis: energy from the sun, 14th International Congress on Photosynthesis. Springer, Dordrecht, pp 339–342

Theiss C, Schmitt F-J, Pieper J, Nganou C, Grehn M, Vitali M, Olliges R, Eichler HJ, Eckert H-J (2011) Excitation energy transfer in intact cells and in the phycobiliprotein antennae of the chlorophyll *d* containing cyanobacterium *Acaryochloris marina*. J Plant Physiol 168(12):1473–1487

Tu J-M (2008) Phycobiliprotein lyases: structure of reconstitution products and mechanistic studies, Dissertation, Munich, Germany

Wang XQ, Li LN, Chang WR, Zhang JP, Gui LL, Guo BJ, Liang DC (2001) Structure of C-phycocyanin from *Spirulina platensis* at 2.2 Å resolution: a novel monoclinic crystal form for phycobiliproteins in phycobilisomes. Acta Crystallogr Sect D 57:784–792

Wang L, Wang S, Fu X, Sun L (2015) Characteristics of an R-Phycoerythrin with two γ subunits prepared from red macroalga *Polysiphonia urceolata*. PLoS One 10(3):e0120333

Watanabe M, Ikeuchi M (2013) Phycobilisome: architecture of a light-harvesting supercomplex. Photosynth Res 116:265–276

Wehrmeyer W (1983) Organization and composition of cyanobacterial and rhodophycean phycobilisomes. In: Papageorgiou GC, Packer L (eds) Photosynthetic prokaryotes: cell differentiation and function. Elsevier Biomedical, New York, pp 1–22

Wilbanks SM, Glazer AN (1993) Rod structure of a phycoerythrin II containing phycobilisome. II. Complete sequence and bilin attachment site of a phycoerythrin gamma subunit. J Biol Chem 268:1236–1241

Wilk KE, Harrop SJ, Jankova L, Edler D, Keenan G, Sharples F, Hiller RG, Curmi PMG (1999) Evolution of a light-harvesting protein by addition of new subunits and rearrangement of conserved elements: crystal structure of a cryptophyte phycoerythrin at 1.63 Å resolution. Proc Natl Acad Sci U S A 96:8901–8906

Wu S-H, Lagarias JC (2000) Defining the bilin lyase domain: lessons from the extended phytochrome superfamily. Biochemistry 39:13487–13495

Yamanaka G, Glazer AN, Williams RC (1980) Molecular architecture of a light-harvesting antenna: comparison of wild type and mutant *Synechococcus* 6301 phycobilisomes. J Biol Chem 255:11004–11010

Yi ZW, Huang H, Kuang TY, Sui SF (2005) Three-dimensional architecture of phycobilisomes from *Nostoc flagelliforme* revealed by single particle electron microscopy. FEBS Lett 579:3569–3573

Zehetmayer P, Hellerer T, Parbel A, Scheer H, Zumbusch A (2002) Spectroscopy of single phycoerythrocyanin monomers: dark state identification and observation of energy transfer heterogeneities. Biophys J 83:407–415

Zhao KH, Su P, Böhm S, Song B, Zhou M, Bubenzer C, Scheer H (2005) Reconstitution of phycobilisome core-membrane linker, LCM, by autocatalytic chromophore binding to ApcE. Biochim Biophys Acta 1706:81–87

Zilinskas BA, Greenwald LJ (1986) Phycobilisome structure and function. Photosynth Res 10:7–35

Gene Manipulation and Biosynthesis of Phycobiliproteins

4

4.1 Introduction

The diverse kinds of tetrapyrrole-containing pigments in many prokaryotic and eukaryotic organisms are classified as bilin pigments. In humans, there are two forms of bilin pigments such as bilirubin and biliverdin IXα which are generated by decomposition of hemoglobin. Apart from humans, the pigment biliverdin IXα is present in a wide variety of animals including bird eggshells, leech cuticles, shells of mollusk, wings of butterfly, butterfly larvae, millipedes, fish fins, skeletons, tissues of spiders, blue corals, certain yeasts, and fungi (Stadnichuk et al. 2015). In leguminous plants, root nodules are pink in color containing leghemoglobin, which also contain chromophores of biliverdin IXα for efficient nitrogen fixation (Ikeuchi and Ishizuka 2008). The phycobiliproteins (PBPs) are a family of accessory light-harvesting protein complexes found in cyanobacteria, red algae, and cryptomonads. PBPs are playing a crucial role in absorption of incident solar radiation for photosynthesis. They are associated with noncyclic linear/open-chain tetrapyrrole prosthetic groups that consist of four pyrrole rings (A, B, C, and D) integrated in α, β, and γ biliproteins. The pyrrole rings of chromophores are designated by capital Latin letters. The two terminal rings are designated as A and D which bear an oxygen atom, while internal rings B and C are bound with adjacent residues (Stadnichuk et al. 2015). Commonly, PBPs are associated with αβ heterodimers in which each subunit carries bilin(s) thioether-linked to particular cysteinyl residues (Sidler 1994; Tooley et al. 2001). Unlike heme and chlorophyll structure, the phycobilins have no content of metal ions. On the basis of chromophore structure, there are four kinds of phycobilins found in cyanobacteria: phycoerythrobilin (PEB), phycocyanobilin (PCB), phycourobilin (PUB), and phycoviolobilin (PVB) (Glazer 1989) (Fig. 4.1).

The composition of PBPs is united together with organization of photometric heterodimers α and β subunits. Moreover, nonpigmented linker polypeptides are embedded in the structure which are responsible for stabilization and modulation of energy absorption within PBPs and help in effective energy transfers toward the photosystem (Glazer 1989). Most of the biliproteins are covalently bound to

V.K. Kannaujiya et al., *Phycobiliproteins: Recent Developments and Future Applications*, https://doi.org/10.1007/978-981-10-6460-9_4

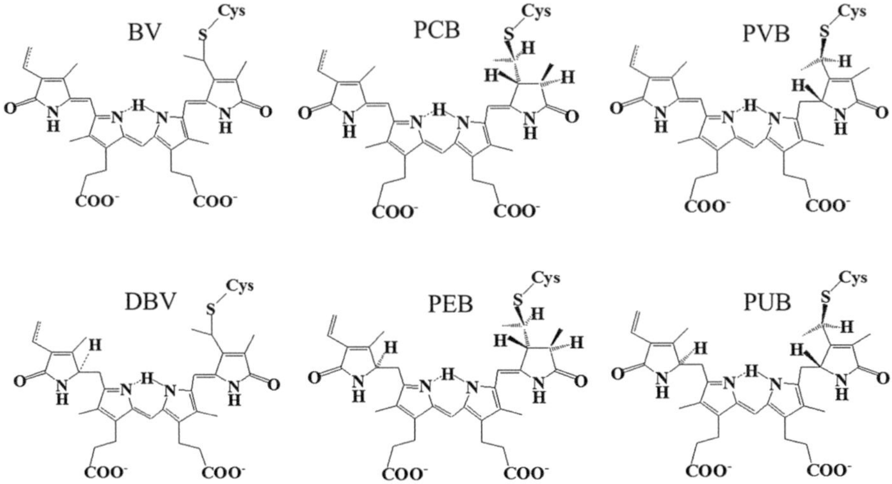

Fig. 4.1 Structure of chromophores in phycobiliproteins. The color molecular structure indicates native color of chromophore attached via A ring by thioether bond on a conserved cysteine residue of bilin proteins. BV (biliverdin), DBV (Fd-dependent bilin reductase), PCB (phycocyanobilin), PEB (phycoerythrobilin), PUB (phycourobilin), and PVB (phycoviolobilin)

cysteine residue by thioether bonds in A ring, while PEB and PUB are covalently bound in D ring (Glazer 1989; Scheer and Zhao 2008). There are several steps involved in biosynthesis of bilin and functional attachment of chromophore (Fairchild and Glazer 1994a; Frankenberg et al. 2001). In order to make a functional accessory light-harvesting complex, there are posttranslational modifications that occur during attachment of biliproteins (PEB and PCB) from chromophores of α and β subunits via thioether linkages at the specific position of cysteine residues (Schluchter et al. 2010). PUB and PVB are derived from isomerization modification by a lyase enzyme of PEB and PCB, respectively (Zhao et al. 2000; Blot et al. 2009). The bilin lyase enzymes have a prominent role in catalyzing the attachment of bilin through desired Cys residues for making functional PBPs (Zhao et al. 2000; Shen et al. 2006; Zhao et al. 2007a; Saunée et al. 2008; Scheer and Zhao 2008). The α subunits of PCB (CpcA) lyase are made up of CpcE/CpcF heterodimers called as the first type of lyase enzyme (Zhou et al. 1992; Swanson et al. 1992; Bhalerao et al. 1994; Biswas et al. 2010). A second family of bilin lyases has been discovered recently and named as T family lyase (Shen et al. 2006). Moreover, S/U family bilin lyase was kept in the third group (Biswas et al. 2010). All bilin lyases required post-translational modification and functionalization of PBPs. The cyanobacterium *Synechococcus* sp. strain PCC 7002 is the most-studied organism for genetic and functional aspect of bilin lyase. Another posttranslational modification occurs through methylation of asparagine at γ-nitrogen position in β subunits of PBPs in cyanobacteria, red algae, and cryptomonads (Rümbeli et al. 1987; Wilbanks et al. 1989; Ducret et al. 1994; Apt et al. 2001). Much progress has been made about biosynthesis of PBPs and enzymes involved in biosynthesis process. This chapter provides the details on studies about the mechanism of biosynthesis of PBPs and probable involvement of genes and enzymes which act as precursors.

4.2 Biosynthesis Mechanism

The biosyntheses of PBPs and tetrapyrrole chromophores are started by a two- to threefold increase in heme pigment (Stadnichuk et al. 2015). PBPs are biosynthesized from heme by the enzymatic catalysis of heme oxygenase (HOs) that converts heme into biliverdin IXα (BV) (Dammeyer and Frankenberg-Dinkel 2008). To complete the catalytic cycle of heme, oxygenase requires three molecules of molecular oxygen and seven reducing electrons which leads to conversion of BV IXα chromophore (Wilks 2002). In addition, the conversion of biliverdin IXα to bilirubin IXα is catalyzed by biliverdin reductase (BvdR) in the presence of reducing cofactors of NADH and NADPH (Sugishima et al. 2005). The enzyme biliverdin reductase is found in several freshwater as well as marine cyanobacteria. These enzymes specifically catalyze the reduction of the methine bridge between ring B and C of BV IXα which leads to conversion into bilirubin IXα (BR IXα) (Overkamp et al. 2014). However, the detailed function of the BR IXα in cyanobacteria is still not identified completely. BR IXα is not incorporated as known light sensors or light-harvesting proteins complex during photosynthesis. There are certain specific genes called FDBR family genes (ferredoxin-dependent bilin reduction) which play an indispensable role in the reduction of bilin to phycobilin in photosynthetic cyanobacteria (Frankenberg et al. 2001). In a current scenario, a group of genes (FDBR family) are identified including *HO* (heme oxygenase), *pebA* (15,16-dihydrobiliverdin:ferredoxin oxidoreductase), *pebB* (phycoerythrobilin:ferredoxin oxidoreductase), *pcyA* (phycocyanobilin:ferredoxin oxidoreductase), and *pebS* (phycoerythrobilin synthase) which play a prominent role in the conversion of BV to phycocyanobilin (PcyA) and phycoerythrobilin (PebA, PebB) (Beale and Cornejo 1991; Dammeyer et al. 2008; Scheer et al. 2015) (Fig. 4.2). Most of the cyanobacteria contain PcyA that is involved in biosynthesis of PCB. The structural configuration of PcyA biliproteins is well understood and revealed an association of α-helix and β-strand sheet. The neutron crystallography data of α-helix sandwich fold shows the presence of highly conserved residues such as Asp105, His88, and His74 which provide polar contacts in substrate BV IXα in conversion of PC (Hagiwara et al. 2006; Unno et al. 2015).

In contrast to PcyA, PebS required four electrons and independently active FDBRs PebA and PebB biliproteins for conversion from BV to PEB including intermediate 15,16-DHBV (Martiny et al. 2006). Structurally, PebS is composed of 233 amino acids which also play an important role in regulation of alternate pathway for the synthesis of PEB (Dammeyer et al. 2008). Some of the intermediate components of harvesting complex are also synthesized in the form of PVB and PUB which alternatively associated with native PBPs. Moreover, the biosynthesis of PVB is originated by molecular isomerization of cysteine-84 residue of PEC in *Mastigocladus laminosus* by catalytic enzymes PecE/PecF lyases (Storf et al. 2001; Böhm et al. 2007).

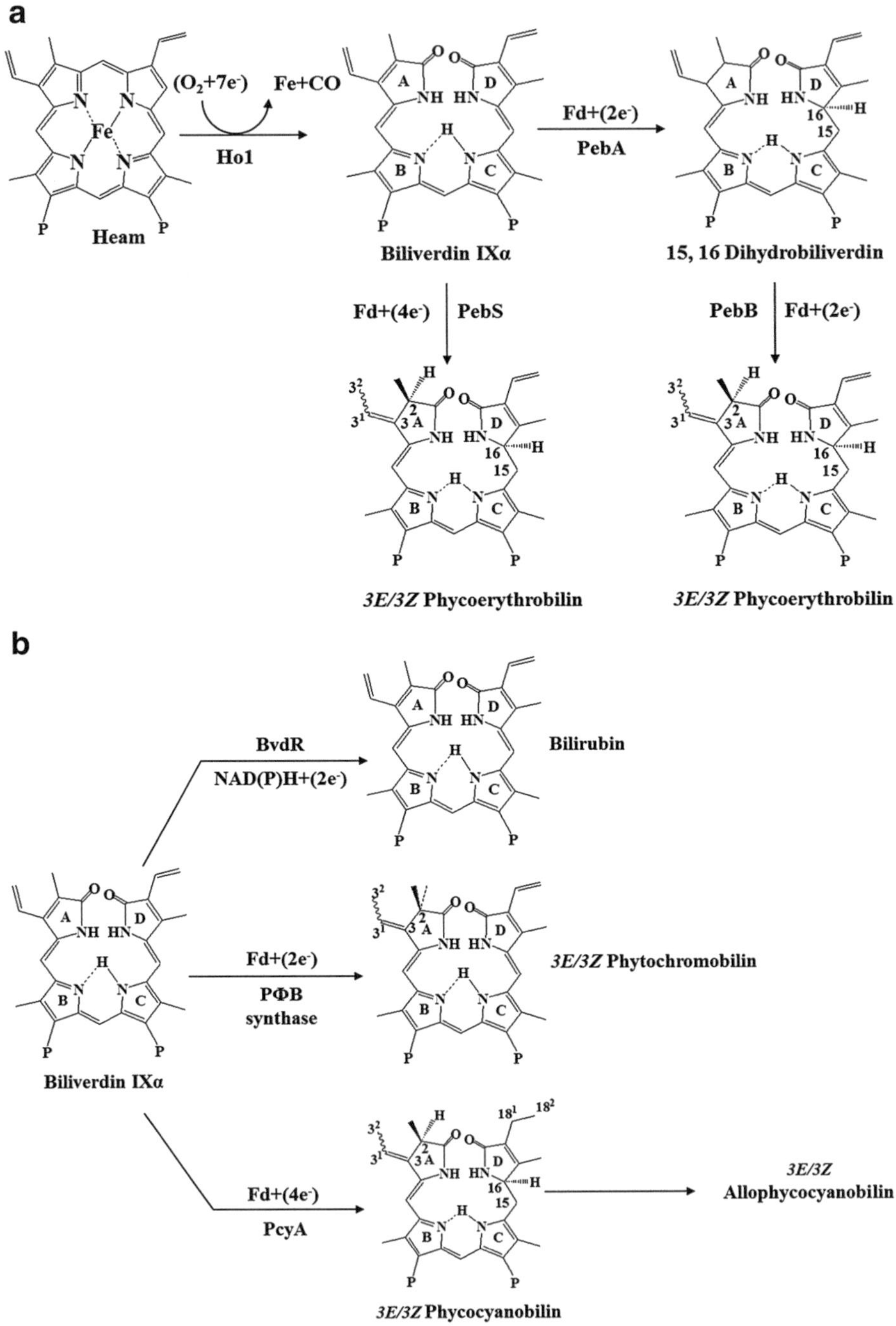

Fig. 4.2 Biosynthesis of chromophores of PE by catalytic reduction electron of biliverdin IXα in the presence of PebA, PebB, and PebS (**a**) and chromophore of PC synthesized by involvement of PcyA, PΦB synthase via the intermediate 18^1,18^2-DHBV from same precursor (**b**) (For details see the references cited in the text)

4.2.1 Posttranslational Modifications of Phycobiliproteins

There are two major types of posttranslational modification pathways that include addition of bilin chromophore in Cys residues and methylation of asparagine residue to make functional β subunits. Bilin lyase enzymes seem to be essential for chromophorylation and ensuring the proper stereochemistry between chromophore and bilin proteins. Prior to attachment of native Cys residues, chromophores required oxidation for specific catalytic attachment to individual Cys residues of biliproteins (Arciero et al. 1988). However, certain cyanobacteria have an ability to change the stereolocation of chromophores in biliproteins via isomerization pathway (Schluchter et al. 2010). The isomerization induces the conversion of bilin proteins to another form, including PVB converted by isomerization of PCB and PEB converted into PUB through isomerization of chromophore by lyase enzyme (Zhao et al. 2000; Blot et al. 2009). In another posttranslational modification, inductive methylation is involved at γ-nitrogen position of asparagine residue in β subunits of PBPs (Ducret et al. 1994; Apt et al. 2001). The methylation of asparagine residues is found in cyanobacteria, red algae, and cryptomonads (Fig. 4.3).

Interestingly, same homologous residue of Asn is found in α subunits which are never modified by methylation; thus the functional role of Asn has not been changed. The site of methylation in Asn residue is very close to the β-82 position bilin chromophore that acts as terminal energy acceptor in trimer; thus it might minimize loss of energy during the energy transfer pathway in bilin proteins (Schluchter et al. 2010; Swanson and Glazer 1990). The mutational analysis reveals that *cpcM* gene is highly sensitive toward light intensity and may be involved in Asn modification on its β subunits (Shen et al. 2008a). In addition, in vitro analysis suggests that Asn modification on its β subunits occurred before formation of trimeric assembly (Miller et al. 2008).

Fig. 4.3 Methylation of the asparagine residue located on β^{72} subunits of PBPs. Reaction catalyzed by methyltransferase (CpcM) using S-adenosylmethionine (SAM) acts as a methyl donor group (Adapted from Clarke 2002; Shen et al. 2008a)

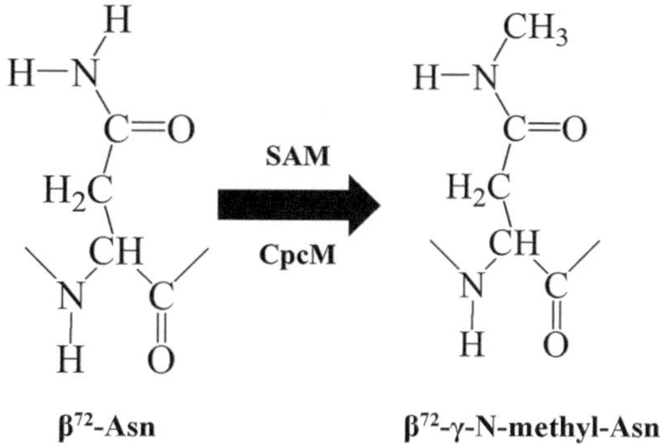

β^{72}-Asn **β^{72}-γ-N-methyl-Asn**

4.2.2 Bilin Attachment

Bilin lyase enzymes are playing an important role in attachment of chromophores to PC and PE bilin precursor. Four types of lyases have been described: (1) E/F-type lyases, (2) S (U)-type lyases, (3) T-type lyases, and (4) Y/Z-type lyase (Biswas et al. 2010) (Table 4.1).

4.2.2.1 CpcE/F-Type Bilin Lyases

The first bilin lyase to be characterized was CpcE/CpcF (Zhou et al. 1992). The ortholog genes of *cpcE* and *cpcF* are located downstream of the *cpcBA* structural genes that encode α and β subunits of PC. The same gene is associated with *cpcC* and *cpcD* gene that encodes PC-associated rod linker polypeptides (Schluchter et al. 2010). The recombinant association of CpcE and CpcF acts as heterodimeric lyase to identify the position of the chromophore attachment at Cys-84 of α-PC (CpcA) (Zhou et al. 1992; Fairchild and Glazer 1994a, b; Schluchter et al. 2010). However, same recombinant association of CpcE/CpcF also catalyzes the reverse enzymatic reaction between holo-α subunit and apo-α subunit by cleaving the thioether bond between the protein and the bilin (Fairchild et al. 1992; Fairchild and Glazer 1994a, b). The association of PecE/PecF including CpeY/CpeZ bilin lyases has a homologous sequence to CpcE/CpcF with diverse PBP-encoding operons (Jung et al. 1995; Kahn et al. 1997; Zhao et al. 2000, 2002; Storf et al. 2001). In cyanobacterium *Synechocystis* sp. PCC 6803, CpcE proteins have multiple HEAT-repeat motifs occurring within the nucleotide sequence (Morimoto et al. 2003). As literature reveals, most of CpcE/CpcF bilin lyases contain about five to six HEAT-repeat

Table 4.1 Types of lyase enzymes for biosynthesis of PBPs

Category	Lyase	Species	Site of action	References
E/F	CpcE/ CpcF	*Synechococcus* sp. PCC 7002	Cys-84-αPC	Fairchild et al. (1992)
E/F	RpcG	*Synechococcus* sp. WH 8102	Cys-84-αPC	Blot et al. (2009)
E/F	PecE/PecF	*Mastigocladus laminosus*	Cys-84-αPEC	Zhao et al. (2005)
T	CpcT	*Synechococcus* sp. PCC 7002	Cys-153-βPC	Shen et al. (2006)
T	CpcT	*Nostoc* sp. *PCC 7120*	Cys155-βPC	Zhao et al. (2007a, b)
S/(U)	CpcS-III	*Nostoc* sp. *PCC 7120*	Cys84-βPC	Zhao et al. (2006, 2007a, b)
			Cys84-βPEC	
			Cys84-αAPC	
			Cys84-βAPC	
S/(U)	CpcS-I/ CpcU	*Synechococcus* sp. PCC 7002	Cys82-βPC	Saunée et al. (2008) and Shen et al. (2008a)
			Cys81-αAPC	
			Cys81-βAPC	

For details, see the following references Biswas et al. (2010), Scheer and Zhao (2008), and Schluchter et al. (2010)

motifs that primarily facilitate protein–protein interaction (Andrade et al. 2001; Takano and Gusella 2002).

4.2.2.2 T-Type Bilin Lyases

The second bilin lyase was characterized as Cpc T which is expressed by *Synechococcus* sp. PCC 7002. The paralog *cpcT* gene was sequenced as a cluster gene of *cpeCDESTR* operon in *Fremyella diplosiphon* (Cobley et al. 2002). Shen et al. (2004) have shown the monomeric form of CpcT recombinant attached at Cys-153 position in β subunit that allows formation of *S*-stereoisomer. Apart from cyanobacteria, the cryptophyte member of *Guillardia theta* encodes lyase enzymes by expression of *cpcT*, which functionally complements with *cpcT* of *Synechocystis* sp. PCC 6803 (Bolte et al. 2008). The distribution of *cpcT* gene sequences strongly suggests a probable role in biosynthesis of PE and attachment to Cys-153 position of cysteine residue (Shen et al. 2006). Recently, cyanophage P-HM1 (*Prochlorococcus* P-HM1) contains auxiliary metabolic gene like Φ*cpeT* homologous to host gene of cyanobacterium *Nostoc* sp. PCC 7120 which encodes T-type lyase and plays a significant role in attachment of linear tetrapyrrole chromophores to Cys-155 of β subunits of PBPs during infection (Gasper et al. 2017).

4.2.2.3 S/U-Type Bilin Lyases

S/U-type bilin lyase was the third family of lyase enzymes that are expressed as CpcS, CpcU, CpcV, CpeS, and CpeU by the paralog gene segment *cpcS-I*, *cpcU*, and *cpcV* (*Synechococcus* sp. PCC 7002). There are different types of CpcS existing in cyanobacteria including CpcS-I (*Synechococcus* sp. PCC7002 and *Synechocystis* sp. PCC6803), CpcS-II (marine *Synechococcus* sp.), and CpcS-III (*Anabaena* sp. PCC7120) which participated in catalyzing PCB attachment to β-PC and β-APC subunits, PC with PEB chromophores, and PCB to Cys-82 in a variety of PBPs, respectively (Ong and Glazer 1988; Shen et al. 2008b; Zhao et al. 2007b). The null mutants of *cpcS-I* and *cpcU* in *Synechococcus* sp. PCC 7002 have decreased the concentration of PC, while Cys-82 chromophores are absent in β-PC. However, certain molecules of PC are still non-covalently bound to PCB chromophore in respective binding pocket (Shen et al. 2008b). The binding association of PCB to Cys-81 chromophore in α and β subunits has been shown during mutant analysis of CpcS-I/CpcU (Shen et al. 2008b). Similarly, the recombinant association of CpcS-I and Cpc U occurs at 1:1 heterodimer that primarily attaches to Cys-82 and Cys-81 of β and α subunits of AP (Saunée et al. 2008). The S/U families of lyases have strong association at equivalent positions of Cys-82 with all PBPs; thus they have broader substrate specificity as compared to E/F-type lyases (Scheer and Zhao 2008). Similarly, cyanobacterium *Nostoc* sp. PCC 7120 also apparently encodes a paralog of CpcS that lacks bilin lyase activity (Zhao et al. 2006, 2007b). The equivalent composition of CpcS and CpcU lyases of *Synechocystis* sp. PCC 6803 is evolved in bilin addition to Cys-82 and forms heterodimer of CpcB subunits (Miller 2007). Recently, CpcS-III is isolated from *T. elongatus* BP1 and solved the structure by X-ray data analysis which reveals that the structure of lyase is dimeric in nature (Kuzin et al. 2007). All of these lyase proteins occur in the form of monomers, homodimers, heterodimers, and tetramers which alternatively bind with ligands such as retinols, fatty acids, pheromones, carotenoids, prostaglandins, and biliverdin (Hieber et al. 2000; Charron et al. 2005; Grzyb et al. 2006).

4.2.2.4 Y/Z-Type Bilin Lyases

The Y/Z-type lyases are known for synthesis of PEI and PEII subunits. These lyases are heterodimeric in nature and catalyze binding reaction between PE α and β subunits (Scheer and Zhao 2008). However, precise site specificity for chromophore binding is still unknown. Recently, an isomerase enzyme Mpe Z has been isolated from marine *Synechococcus* sp. that catalyzes the binding reaction between chromophore and Cys-83 position in PEII of α subunit. In the same reaction, PEB is also converted into PUB and has shown chromatic shifting from green to blue spectrum (Shukla et al. 2012; Singh et al. 2015). The core–membrane linker protein (L_{CM}^{99}), also called ApcE, is found to have a unique type of autocatalytic bilin addition activity (Zhao et al. 2005). This ApcE domain exhibits intrinsic bilin lyase activity for attaching PCB in an appropriate conformation to yield a red-shifted PCB product from 662 to 672 nm of wavelength (Biswas et al. 2010).

4.3 Gene Manipulation and Transformation

The strains of *Escherichia coli* are serving as a model organism for investigation of biosynthesis pathways of PBPs. Several models have been proposed for biosynthesis of PBPs (Fig. 4.4).

The synthesis mechanism of PC and PEC holo-α subunits was first reconstituted in heterologous host of *E. coli* by using a dual plasmid system (Tooley et al. 2001; Tooley and Glazer 2002). PCB synthesis was induced by heterologous gene expression in *E. coli* including co-expression of PCB-ferredoxin oxidoreductase (*pcyA*) and heme oxygenase 1 (*hox1*) (Landgraf et al. 2001). The holo subunits of PCB were synthesized by co-expression of *cpcE/F*, *hox1*, *cpcA*, and *pcyA* genes which involves biosynthesis of holo-α-PC (Tooley et al. 2001). Biswas et al. (2010) recognize apo-PC subunit expression in *E. coli* for effective quantity and high fluorescence intensity for binding of recombinant PCB in cyanobacteria. A similar report has been found for α and β subunits of APC (ApcA and ApcB) in same organism (Zhao et al. 2007b). Since the heterologous system is largely unknown, thus expression of single subunits of bilin proteins was analyzed (Arciero et al. 1988; Liu et al. 2010). The APC trimers as core subunits of bilin proteins exhibit holo-ApcA and ApcB assembly in *E. coli* which is primarily found in *Synechocystis* sp. PCC 6803. The binding association from PCB to APC may be induced during initiation of trimeric assembly (Liu et al. 2010). The homologous expressions of gene expression of PBPs are still to be discovered for productive biosynthesis and functional properties in *E coli.* system. The tools of gene manipulation and recombinant production of PBPs substantially reduce the market cost.

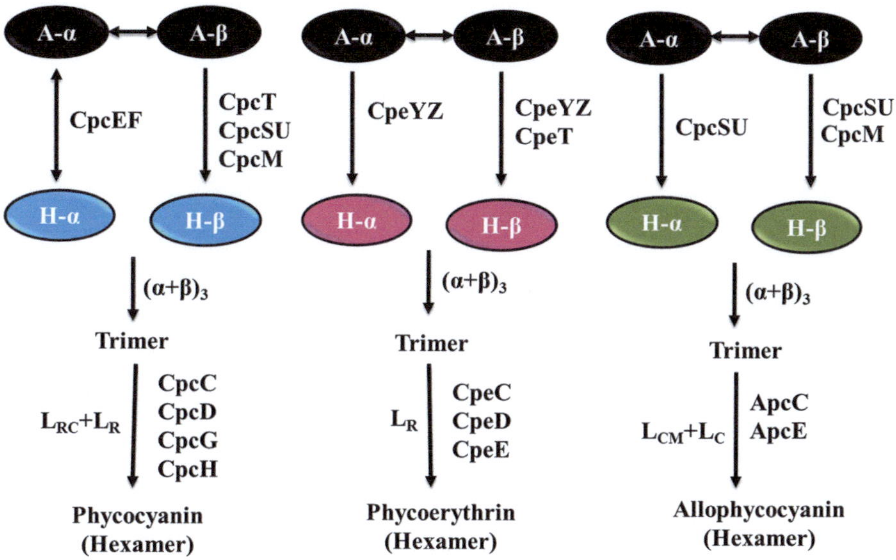

Fig. 4.4 A model for the biosynthesis of rod and core subunits of PBPs (PC, PE, and APC) through catalysis of lyase enzymes in cyanobacteria (For details see the references cited in the text)

4.4 Chromatic Adaptation

Certain cyanobacteria have a distinguished ability to alter pigment composition in response to changes in wavelength of light regimes for adaptation and sustainable growth in environment condition (Postius et al. 2001). This phenomenon is called complementary chromatic acclimation (CCA), where PCB is converted into PEB in red light whereas PEB is converted into PCB under green light in *Fremyella diplosiphon* (Bennett and Bogorad 1973; Kehoe and Gutu 2006; Gutu and Kehoe 2012; Airs et al. 2014; Gan et al. 2014; Gan and Bryant 2015; Zhao et al. 2015) (Fig. 4.5). CCA has also been recognized for photomorphogenesis in cyanobacteria where there is reformation of morphology upon exposure of different spectra of light (Singh et al. 2015). There are four types of CCA reported in cyanobacterial species which contains a combination of PC and PE (Tandeau de Marsac 1977). The complementary chromatic adaptation occurred mostly in freshwater cyanobacteria such as *Fremyella diplosiphon* (Kehoe and Gutu 2006), *Calothrix* sp. PCC 7601 (Kehoe and Gutu 2006), *Tolypothrix tenuis* (Ohki and Fujita 1978), and *Nostoc punctiforme* (Hirose et al. 2010).

Tendeau de Marsac (1977) had described four groups such as group I, II, III, and IV embedded in different strains of cyanobacteria. In type I CCA, a group of organisms does not have the ability to convert PC to PE or PE to PC upon exposure of different wavelengths of light. In type II CCA, only PE levels have been changed in response to green and red spectrum of light wavelengths, whereas type III CCA exhibits proper changes in PE and PC upon external stimulus of green or red light

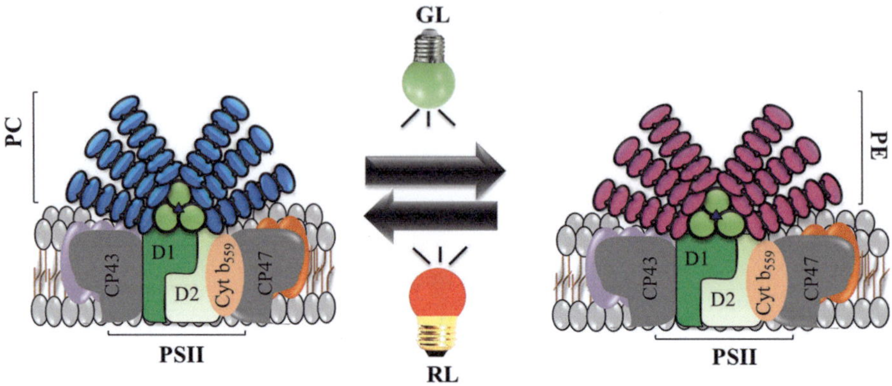

Fig. 4.5 Light-induced complementary chromatic adaptation and conversion of PC to PE or PE to PC. GL green light, RL red light, PSII photosystem II

wavelength (Tandeau de Marsac 1977; Gutu and Kehoe 2012). In type IV CCA, marine strains of cyanobacteria exhibit changes in blue and green wavelength by alteration of chromophore association through bilin proteins (Everroad et al. 2006; Gutu and Kehoe 2012). Recently, in Gan et al. (2015) a new type of CCA was identified in *Leptolyngbya* sp. strain JSC-1 that synthesized red-shifted chlorophyll between Chl *d* and Chl *f* grown under far-red light (FRL). There was integration of a 21-photosynthetic gene cluster regulated by a two-component system (*rfpA*/*B*/*C*) expressed under FRL (Gan et al. 2015). Very recently, it has demonstrated a novel CCA in *Halomicronema hongdechloris* that shows loss of PC and remodeling of the APC core subunits with novel FRL-absorbing components (Li et al. 2016). The red-shifted PBPs represent a novel strategy for CCA that provides a new challenge for global needs under agriculture production. Similarly, photosensory proteins have also shown a physiological response under a wide wavelength of light in many cyanobacteria (Montgomery 2016). Certain cyanobacterial photosensory proteins have shown red/far-red sensitivity like higher plants including *F. diplosiphon* (Quest et al. 2007) and thermophilic *Synechococcus* sp. OS-A (Ulijasz et al. 2008). RfpA is another class of photosensory protein that responds to red and far-red light (Gan et al. 2014). In addition, cyanobacteria have extensive photosensory regulation for red-/green-responsive proteins including CcaS from *Synechocystis* sp. PCC 6803 and *Nostoc punctiforme* ATCC 29133, AnPixJ from *Nostoc* sp. PCC7120, and NpR3784 from *Nostoc punctiforme* (Hirose et al. 2010; Narikawa et al. 2008, 2015; Rockwell et al. 2015). Similarly, a blue-/green-responsive sensory domain has also been found in cyanobacteria including TePixJ from *Thermosynechococcus elongatus* and PixJ/TaxD1 from *Synechocystis* sp. (Ishizuka et al. 2006; Yoshihara et al. 2006). Certain complex cyanobacteriochrome has membrane embedded with seven GAF domains and a progressive response under different spectra of wavelength (Gottlieb et al. 2015; Rockwell et al. 2012). Recently, through a cyanobacterium *Microcoleus* IPPAS B353, it has been identified that cyanobacteriochromes adapted in high light near-UV/violet spectrum which indicated a distinct feature of blue-light receptor (Narikawa et al. 2006; Okajima et al. 2006; Cho et al. 2015). Several genes have been reported until now for a probable role in chromatic adaptation and morphogenesis in few cyanobacteria (Table 4.2).

4.4.1 PE–PC Photoconversion

A few cyanobacterial species are able to change their protein–pigment ratio under varying wavelengths of light for enhancement of photosynthetic productivity. The composition of PBPs is the most remarkable example for acclimation under different wavelengths of light in an open environment. To sustain the dynamic and fluctuating environmental condition, photosynthetic organisms adapted by regulation of filament color and cell shape (Singh and Montgomery 2011; Montgomery 2015), PBPs (Tandeau de Marsac 1977), chlorophyll content (Osborne and Raven 1986) and several photoprotective mechanisms (Kirilovsky and Kerfeld 2012). The consequent alteration in change of phenotypic color of cyanobacterial filaments is a primary response for light-dependent acclimation or complementary chromatic adaptation (Bogorad 1975). Indeed, cyanobacteria sense green (GL) and red (RL) color wavelength of light in an open environment and simultaneously adapt through interconversion in PC and PE subunits of PBPs. However, cyanobacterial filaments exhibit synthesis of more rod cells when grown in low light as compared to high-light condition (Grossman et al. 1986). Interestingly, relative subunits of PEI and PEII are not able to change in color due to association of two chromophores (PUB and PEB) under blue and white light (Everroad et al. 2006). Agostoni et al. (2016) found rapid accumulation of PBP rods under inconsistent light condition that induces more responsiveness and fitness for photosystem in *F. diplosiphon.* The photo-reversible nature of pigments in cyanobacteria exhibits unique plasticity to optimize their growth and development under various light qualities. In the present scenario, CCA-induced photoregulation has been a focus in the study of photoreceptors and associated signaling transduction pathways involved in optimization of photosynthetic efficiency and photomorphogenesis in prokaryotic organisms (Montgomery 2008).

4.4.2 Signal Transduction

4.4.2.1 RcaC–RcaE Regulation

Genetically, chromatic adaptation is regulated by a phytochrome-like protein called RcaE, first found in prokaryotic system (Kehoe and Grossman 1996; Wiltbank and Kehoe 2016). RcaE is a unique class of photoregulated protein which contains a photosensor domain that covalently binds to tetrapyrrole chromophore (Rockwell and Lagarias 2010). RcaE regulates the phosphorelay cascade pathway between RcaF and RcaC which primarily regulate the transcription of biosynthetic genes for PBPs (Alvey et al. 2007; Li et al. 2008; Bezy and Kehoe 2010). *Fremyella diplosiphon* is a model organism for the RcaE mechanism which controls the response in

Table 4.2 Photoreceptor for pigmentation and morphological alteration in cyanobacteria in response to light and nutrient stress

Organism	Photoreceptor (accession no.)	Light color	Modulators and effectors	Physiological control	References
Fremyella diplosiphon	RcaE (GB: U59741)	R/G	RcaF, RcaC, CpeR	Pigmentation, morphology	Li et al. (2008) and Pattanaik et al. (2011)
			BolA	Morphology, ROS	Singh and Montgomery (2014, 2015)
			MreB, MreC, MreD	Morphology	Singh and Montgomery (2014)
			TspO	Stress, morphology	Busch and Montgomery (2015)
	IflAc (GB: KF316936)	B/G/R/FR	ND	Growth at low cell density	Bussell and Kehoe (2013)
	DpxA (WP_045867496)	T/Y	ND	Pigmentation	Wiltbank and Kehoe (2016)
Synechocystis sp. PCC 6803	PixJ/TaxD1 (*sll0041*)	B/G	ND	Phototaxis	Yoshihara et al. (2006)
	PixA/ UirS (*slr1212*)	UV-A/B/G	NixB/UirR, NixC/ LsiR	Phototaxis	Narikawa et al. (2011) and Song et al. (2011)
	Cph1 (*slr0473*)	R/FR	Rcp1	Growth	Fiedler et al. (2004)
	Cph2f (*sll0821*)	B/G	Intragenic effector	Phototaxis, growth	Moon et al. (2011) and Savakis et al. (2012)
	CcaS (*sll1473–5*)	R/G	CcaR	Pigmentation	Hirose et al. (2013)
	PlpA (*sll1124*)	B	ND	Growth	Wilde et al. (1997)

Nostoc punctiforme ATCC 29133	CcaS(Npun_F3797)	R/G	CcaR	Pigmentation	Hirose et al. (2010)
	PtxD (NpF2164)	Wg	ND	Phototaxis	Campbell et al. (2015)
Chroococcidiopsis thermalis PCC 7203	RfpA (7203:Chro_4230)	R/FR	RfpB, RfpC	Pigmentation	Zhao et al. (2015)
Anabaena sp./*Anabaena* sp. PCC 7120	AphC (*all2699*)	R/FR	CyaC	Motility	Ohmori et al. (2002) and Okamoto et al. (2004)
Thermosynechococcus elongatus	SesAf (*tlr0924*)	B/G	Intragenic domain	Cellular aggregation	Enomoto et al. (2015)
	SesBh (*tlr1999*)	B/T	Intragenic domain	Cellular aggregation	Enomoto et al. (2015)
	SesCf, h (*tlr0911*)	B/G	Intragenic domain	Cellular aggregation	Enomoto et al. (2015)

Adapted from Montgomery (2016)

both red and green light (Montgomery 2007). In CCA adaptation, PBS shows dramatically reversible change in the phenotype of cells, from blue-green to brick-red in color under red and green light, respectively (Kehoe and Gutu 2006). RcaE exhibits kinase activity in red light (Hirose et al. 2013) which initiates electron phosphorelay cascade pathway between two response regulators, such as RcaF and RcaC (DNA-binding domain), which are photoregulated proteins (Grossman and Kehoe 1997). The DNA-binding domain of RcaC is regulated by the transcription pathway of PE and PC encoding genes that play a crucial role in CCA (Li et al. 2008; Bezy and Kehoe 2010; Gutu and Kehoe 2012) (Fig. 4.6).

RcaE is linked with the histidine kinase domain that forms interconvertible switches in green and red spectrum such as kinase-driven green-absorbing form ($RcaE^G$) in red light and kinase-driven red-absorbing form ($RcaE^R$) in green light (Hirose et al. 2013). In red light, the phosphorylation occurs in $RcaE^G$ complex in a two-component system with single-domain response regulator RcaF and transcription factor RcaC (Wiltbank and Kehoe 2016). The *cpeCDESTR* and *cpeBA* operon systems are required to produce PE-containing PBPs. However, operon *cpcB2A2* transcription is required to produce PC of PBPs (Chiang et al. 1992). Notably, the *F. diplosiphon* genome has single copies of *cpeC* and *cpeBA* operon for regulation of photoregulation (Kehoe and Gutu 2006). Indeed, it was recently reported that *F. diplosiphon* has 27 putative phytochrome-related photoreceptors and 305 potential two-component signaling proteins (Yerrapragada et al. 2015). However, RcaE is suggested to reduce the expression of PE under red light or induce that expression under green light in both *rcaE* and *rcaC* mutants that indicate another pathway for induction of PE genes (Li et al. 2008; Li and Kehoe 2005; Seib and Kehoe 2002) (Fig. 4.7).

Molecular pathway for complex chromatic adaptation in green light (GL) and red light (RL) has been well documented. The expression patterns of RcaE-F-C are

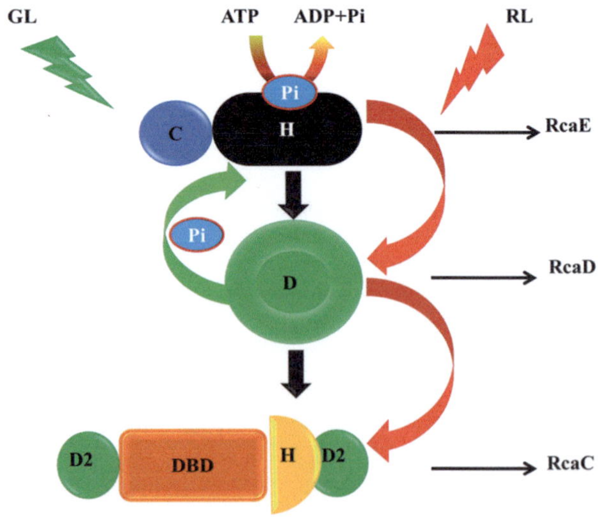

Fig. 4.6 Rca-induced pathway for CCA regulation by absorption of green light (GL) and red light (RL). *C* cysteine residue, *H* histidine residue, *D* aspartate residue, *DBD* DNA-binding domain, *P* phosphate group (For details, see text)

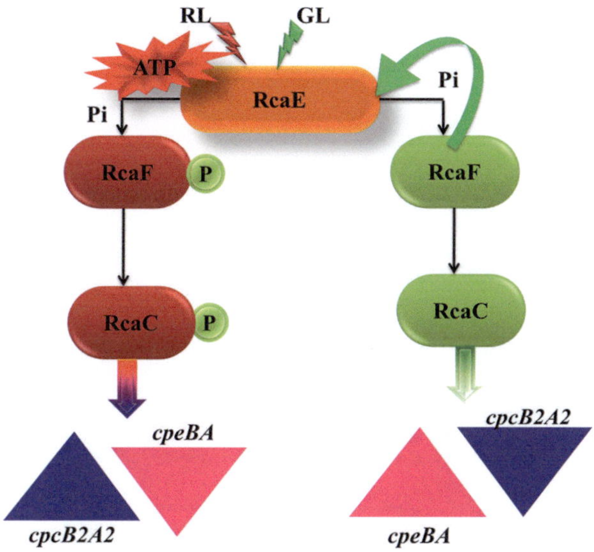

Fig. 4.7 Genetic pathway for CCA regulation under green light (GL) and red light (RL). The expression patterns of RcaE-F-C regulated upon color of light for PC and PE gene expression. Pink color triangle indicates up-/downregulation of PE (*cpeBA*) and blue triangle indicates for PC (*cpcB2A2*) regulation (For details see the references cited in the text)

activated by phosphorylation under red light condition. RcaC-P inhibited expression of *cpeA*, *cpeB*, *cpeC*, *cpeD*, *cpeE*, and *FdTonB* which leads to complete inhibition of PE. Alternatively, RcaC-P promotes upregulation of *pcyA*, *cpcB2*, and *cpcA2* which results to an enhanced PC expression. Under green light, RcaF-P dephosphorylates and stops effector signal response by RcaC. RcaC induces upregulation for expression of *cpeA*, *cpeB*, *cpeC*, *cpeD*, *cpeE*, and *FdTonB* genes, whereas it induces downregulation for expression of PC-related genes such as *cpcB2* and *cpcA2*.

4.4.2.2 CcaS–CcaR Regulation

Recently, the CcaS system, a kind of GAF domain, was discovered in *Synechocystis* sp. PCC 6803 (Hirose et al. 2008). Moreover, the GAF domain of CcaS is homologous to the RcaE domain and functionally follows a similar mechanism. However, CcaS has more photophosphorylation activity in green light as compared to RcaE (Hirose et al. 2008). In addition, CcaS directly phosphorylates a response regulator CcaR which clearly indicates differences from RcaE components related to phosphorylation in RcaF and RcaC (Hirose et al. 2008, 2010) (Fig. 4.8). This has proved that RcaE is able to regulate PBP-encoding genes including CpeR (Cobley et al. 2002) and morphogenes that alter the morphological structure of cyanobacteria under different regimes of light (Bordowitz and Montgomery 2008; Singh and Montgomery 2015).

Nostoc punctiforme ATCC 29133 also regulates the PE synthesis in green and red light by performing group II chromatic adaptation (Tandeau de Marsac 1977). Mutational studies have shown that green light induces the phosphorylation of CcaS

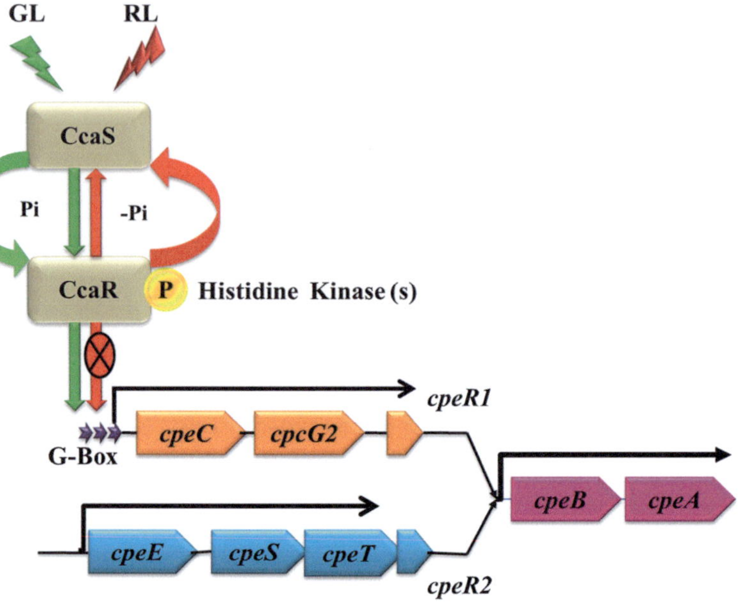

Fig. 4.8 Complementary chromatic adaptation in *Nostoc punctiforme* under green and red light (Adapted from Hirose et al. 2010)

domains into CcaR and phosphorylated CcaR induces *cpeC-cpcG2-cpeR1* operon expression that leads to accumulation of PE. In contrast to green light, CcaS probably represses the activity of cpeC-*cpcG2-cpeR1* by dephosphorylation of CcaR under red light (Hirose et al. 2010). Recently, IflA proteins were identified in *F. diplosiphon* having two distinct photosensory domains which respond in blue, green, red, and far-red light under control of the RcaE domain (Bussell and Kehoe 2013).

The red and far-red light photonic environments are playing a distinguishable role in photoresponse. DpxA is another photoregulator that inhibit synthesis of PE in *F. diplosiphon* (Wiltbank and Kehoe 2016). However, DpxA polypeptide shows an accumulation of PE in yellow light (Wiltbank and Kehoe 2016). The molecular mechanism related to structure and light-induced signaling pathways in photosystem are well described. However, the structural alterations in morphology of cell are not well understood.

4.4.2.3 Photomorphogenesis

The changes of morphology of cyanobacteria are largely independent from CCA photoregulation in different wavelengths of light (Bordowitz et al. 2010; Pattanaik et al. 2012; Walters et al. 2013). RcaE is well demonstrated for regulation in cellular morphology in light-dependent processes (Singh and Montgomery 2015). RcaE is having dual response in quality and quantity of light for morphological alteration response in cyanobacteria under depth of water column (Bordowitz et al. 2010;

Pattanaik et al. 2012; Walters et al. 2013). Several morphogenes have been identified in the study of photo-dependent morphological changes in *F. diplosiphon* (Montgomery 2016). TonB is a transcription factor that upregulated upon exposure of green light and progressively altered cell width (Pattanaik and Montgomery 2010). Moreover, upregulation of TonB was independent; thus, it was not regulated by RcaE. Recently, it has been reported that RcaE induces expression of BolA under red light which is remarkably associated for adaptation of spherical morphology in cyanobacteria (Singh and Montgomery 2015). Moreover, BolA binds to another morphogene and inhibits expression of bacterial actin encoding *mreB* gene in red light (Singh and Montgomery 2014). In green light, the lower concentrations of BolA are allowed for expression of *mreB* gene which gives a rod-shaped morphology to cyanobacteria (Singh and Montgomery 2014, 2015). It has been recently shown that regulation of BolA levels is also altered by reactive oxygen species produced by the cellular system (Singh and Montgomery 2015). The central regulator CpeR is another factor which is involved in PE synthesis under exposure of green light (Seib and Kehoe 2002) and has been shown to have an impact on cellular morphology as introduced above (Pattanaik et al. 2011). The mutant of Δ*cpeR* exhibited rounded cells as compared to wild-type cells under exposure of green light (Pattanaik et al. 2011).

4.5 Second Messenger and Signal Transduction

Second messengers are active molecules or ions which triggered relay signals to effector molecules after receiving from primary messengers on cell-surface receptors. The secondary messengers have amplifying environmental signals received from primary messengers (light) in cyanobacteria (Agostoni and Montgomery 2014). In the present scenario, a wide variety of secondary messengers has been reported in cyanobacteria including calcium (Ca^{2+}), cyclic AMP (cAMP), and c-di-GMP (Montgomery 2016) which is associated with light-regulated photosensory receptors and signaling. Ca^{2+} is the primary element that is triggered by different light conditions in various cyanobacteria and also serves as signal generator during electron transport in processes of photosynthesis (Torrecilla et al. 2004). In response to UV radiation, Ca^{2+} ions are accumulated in heterocysts of cyanobacteria and facilitate adaptive response against damage caused by UV radiation in *Anabaena* sp. (Richter et al. 1999). Another secondary messenger is cyclic AMP (cAMP) found in cyanobacteria that exhibits alteration in concentration upon exposure of light (Terauchi and Ohmori 2004; Raffelberg et al. 2013). A photosensory protein AphC is an acceptor of red and far-red light which mediates to increase cAMP levels in *Anabaena* cells (Okamoto et al. 2004). Recently, it has been shown that iron deprivation also altered *aphC* expression (González et al. 2014). Oxidative stress was another factor for CyaC-dependent intracellular regulation cAMP (Higo et al. 2008). Thus, perception of light may serve as regulation of cAMP upon alteration of nutrient and oxidative stress. Moreover, light-dependent regulation of cAMP is generating signal for regulation of cellular motility in *Synechocystis* (Terauchi and Ohmori 2004). The cyclic di-GMP has been found in several cyanobacteriochromes and controls cellular aggregation, motility, cellular buoyancy, and biofilm formation as secondary messenger (Savakis et al. 2012; Enomoto et al. 2015; Agostoni et al. 2016; Montgomery 2016). Recently, bioinformatics analyses show that light-responsive domains are integrated with domain of c-di.GMP for regulation of intracellular homeostasis (Agostoni et al. 2013; Agostoni and Montgomery 2014). The spectrum of high light was an important factor for c-di. GMP-induced regulation of cellular morphology and differentiation in *Anabaena* sp. (Neunuebel and Golden 2008). ROS is the most important secondary messenger in signal transduction having both potential as signal and damaging agent. The generation of ROS is light dependent, and accumulation is increased under red light condition (Walters et al. 2013). The elevated levels of ROS in *F. diplosiphon* may reduce the filament size with spherical morphology under red light condition (Singh and Montgomery 2012). Moreover, the increase in threshold levels of ROS shows a detrimental effect on cellular growth and survivability of cyanobacteria (Singh and Montgomery 2012). Recently, tryptophan-rich sensory protein (TSPO) was recognized as homolog for tetrapyrrole-binding proteins under regulation of RcaE in *F. diplosiphon* (Busch and Montgomery 2015). However, TSPO-induced response of photoregulation is still to be discovered.

4.6 Conclusion

The current advancement in CCA regulation in cyanobacterial species has begun to resolve strategies of dynamic acclimations in different light regimes. The capacity of CCA adaptation in cyanobacteria makes a global contribution in fixation of carbon and utilization of light quality in the process of photosynthesis. Understanding the biosynthetic pathway of PBPs may fulfill the limitation and contribute to the global economy. Knowledge about PBPs may serve as a key point to improve engineering and biotechnological processes for critical adaptation and biosynthesis against a changing environment.

References

Agostoni M, Montgomery BL (2014) Survival strategies in the aquatic and terrestrial world: the impact of second messengers on cyanobacterial processes. Life 4:745–769

Agostoni M, Koestler BJ, Waters CM, Williams BL, Montgomery BL (2013) Occurrence of cyclic di-GMP-modulating output domains in cyanobacteria: an illuminating perspective. MBio 4:e00451–e00413

Agostoni M, Lucker BF, Smith MAY, Kanazawa A, Blanchard GJ, Kramer DM, Montgomery BL (2016) Competition-based phenotyping reveals a fitness cost for maintaining phycobilisomes under fluctuating light in the cyanobacterium Fremyella diplosiphon. Algal Res 15:110–119

Airs R, Temperton B, Sambles C, Farnham G, Skill S, Llewellyn C (2014) Chlorophyll f and chlorophyll d are produced in the cyanobacterium *Chlorogloeopsis fritschii* when cultured under natural light and near-infrared radiation. FEBS Lett 588:3770–3777

Alvey RM, Bezy RP, Frankenberg-Dinkel N, Kehoe DM (2007) A light regulated OmpR-class promoter element co-ordinates light-harvesting protein and chromophore biosynthetic enzyme gene expression. Mol Microbiol 64:319–332

Andrade MA, Petosa C, O'Donoghue SI, Müller CW, Bork P (2001) Comparison of ARM and HEAT protein repeats. J Mol Biol 309:1–18

Apt KE, Metzner S, Grossman AR (2001) The γ subunits of phycoerythrin from a red alga: position in phycobilisomes and sequence characterization. J Phycol 37:64–70

Arciero DM, Bryant DA, Glazer AN (1988) In vitro attachment of bilins to apophycocyanin I: specific covalent adduct formation at cysteinyl residues involved in phycocyanobilin binding in C-phycocyanin. J Biol Chem 263:18343–18349

Beale SI, Cornejo J (1991) Biosynthesis of phycobilins. Ferredoxin-mediated reduction of biliverdin catalyzed by extracts of *Cyanidium caldarium*. J Biol Chem 266:22328–22332

Bennett A, Bogorad L (1973) Complementary chromatic adaptation in a filamentous blue-green alga. J Cell Biol 58:419–435

Bezy RP, Kehoe DM (2010) Functional characterization of a cyanobacterial OmpR/PhoB class transcription factor binding site controlling light color responses. J Bacteriol 192:5923–5933

Bhalerao RP, Kind LK, Gustafsson P (1994) Cloning of the *cpcE* and *cpcF* genes from *Synechococcus* sp. PCC 6301 and their inactivation in *Synechococcus* sp. PCC 7942. Plant Mol Biol 26:313–326

Biswas A, Vasquez YM, Dragomani TM, Kronfel ML, Williams SR, Alvey RM, Bryant DA, Schluchter WM (2010) Biosynthesis of cyanobacterial phycobiliproteins in *Escherichia coli*: chromophorylation efficiency and specificity of all bilin lyases from *Synechococcus* sp. strain PCC 7002. Appl Environ Microbiol 76(9):2729–2739

Blot N, Wu XJ, Thomas JC, Zhang J, Garczarek L, Bohm S, Tu JM, Zhou M, Ploscher M, Eichacker L, Partensky F, Scheer H, Zhao KH (2009) Phycourobilin in trichromatic phycocyanin from oceanic cyanobacteria is formed post-transnationally by a phycoerythrobilin lyase-isomerase. J Biol Chem 284:9290–9298

Bogorad L (1975) Phycobiliproteins and complementary chromatic adaptation. Annu Rev Plant Physiol 26:369–401

Böhm SB, Endres S, Scheer H, Zhao KH (2007) Biliprotein chromophore attachment: chaperone-like function of the PecE subunit of alpha-phycoerythrocyanin lyase. J Biol Chem 282:25357–25366

Bolte K, Kawach O, Prechtl J, Gruenheit N, Nyalwidhe J, Maier UG (2008) Complementation of a phycocyanin-bilin lyase from *Synechocystis* sp PCC 6803 with a nucleomorph-encoded open reading frame from the cryptophyte *Guillardia theta*. BMC Plant Biol 8:56

Bordowitz JR, Montgomery BL (2008) Photoregulation of cellular morphology during complementary chromatic adaptation requires sensor kinase-class protein RcaE in *Fremyella diplosiphon*. J Bacteriol 190:4069–4074

Bordowitz JR, Whitaker MJ, Montgomery BL (2010) Independence and interdependence of the photoregulation of pigmentation and development in *Fremyella diplosiphon*. Commun Integr Biol 3:151–153

Busch AWU, Montgomery BL (2015) The tryptophan-rich sensory protein (TSPO) is involved in stress-related and light-dependent processes in the cyanobacterium *Fremyella diplosiphon*. Front Microbiol 6:1393

Bussell AN, Kehoe DM (2013) Control of a four-color sensing photoreceptor by a two-color sensing photoreceptor reveals complex light regulation in cyanobacteria. Proc Natl Acad Sci USA 110:12834–12839

Campbell EL, Hagen KD, Chen R, Risser DD, Ferreira DP, Meeks JC (2015) Genetic analysis reveals the identity of the photoreceptor for phototaxis in hormogonium filaments of *Nostoc punctiforme*. J Bacteriol 197:782–791

Charron JBF, Ouellet F, Pelletier M, Danyluk J, Chauve C, Sarhan F (2005) Identification, expression, and evolutionary analyses of plant lipocalins. Plant Physiol 139:2017–2028

Chiang GG, Schaefer MR, Grossman AR (1992) Complementation of a red-light-indifferent cyanobacterial mutant. Proc Natl Acad Sci USA 89:9415–9419

Cho SM, Jeoung SC, Song J-Y et al (2015) Genomic survey and biochemical analysis of recombinant candidate cyanobacteriochromes reveals enrichment for near UV/Violet sensors in the halotolerant and alkaliphilic cyanobacterium *Microcoleus* IPPAS B353. J Biol Chem 290:28502–28514

Clarke S (2002) The methylator meets the terminator. Proc Natl Acad Sci USA 99:1104–1106

Cobley JG, Clark AC, Weerasurya S, Queseda FA, Xiao JY, Bandrapali N, D'Silva I, Thounaojam M, Oda JF, Sumiyoshi T, Chu MH (2002) CpeR is an activator required for expression of the phycoerythrin operon (*cpeBA*) in the cyanobacterium *Fremyella diplosiphon* and is encoded in the phycoerythrin linker- polypeptide operon (*cpeCDESTR*). Mol Microbiol 44:1517–1531

Dammeyer T, Frankenberg-Dinkel N (2008) Function and distribution of bilin biosynthesis enzymes in photosynthetic organisms. Photochem Photobiol Sci 7:1121–1131

Dammeyer T, Bagby SC, Sullivan MB, Chisholm SW, Frankenberg-Dinkel N (2008) Efficient phage-mediated pigment biosynthesis in oceanic cyanobacteria. Curr Biol 18:442–448

Ducret A, Sidler W, Frank G, Zuber H (1994) The complete amino acid sequence of R phycocyanin-I α and β subunits from the red alga *Porphyridium cruentum*: structural and phylogenetic relationships of the phycocyanins within the phycobiliproteins families. Eur J Biochem 221:563–580

Enomoto G, Ni-Ni-Win NR, Ikeuchi M (2015) Three cyanobacteriochromes work together to form a light color-sensitive input system for c-di-GMP signaling of cell aggregation. Proc Natl Acad Sci USA 112:8082–8087

Everroad C, Six C, Partensky F, Thomas J-C, Holtzendorff J, Wood AM (2006) Biochemical bases of type IV chromatic adaptation in marine *Synechococcus* spp. J Bacteriol 188:3345–3356

Fairchild CD, Glazer AN (1994a) Non-enzymatic bilin addition to the α subunit of an apophycoerythrin. J Biol Chem 269:28988–28996

Fairchild CD, Glazer AN (1994b) Oligomeric structure, enzyme kinetics, and substrate specificity of the phycocyanin alpha subunit phycocyanobilin lyase. J Biol Chem 269:8686–8694

Fairchild CD, Zhao J, Zhou J, Colson SE, Bryant DA, Glazer AN (1992) Phycocyanin α subunit phycocyanobilin lyase. Proc Natl Acad Sci USA 89:7017–7021

Fiedler B, Broc D, Schubert H, Rediger A, Börner T, Wilde A (2004) Involvement of cyanobacterial phytochromes in growth under different light qualities and quantities. Photochem Photobiol 79:551–555

Frankenberg N, Mukougawa K, Kohchi T, Lagarias JC (2001) Functional genomic analysis of the HY2 family of ferredoxin-dependent bilin reductases from oxygenic photosynthetic organisms. Plant Cell 13:965–972

Gan F, Bryant DA (2015) Adaptive and acclimative responses of cyanobacteria to far-red light. Environ Microbiol 17:3450–3465

Gan F, Zhang S, Rockwell NC, Martin SS, Lagarias JC, Bryant DA (2014) Extensive remodeling of a cyanobacterial photosynthetic apparatus in far-red light. Science 345:1312–1317

Gan F, Shen G, Bryant DA (2015) Occurrence of far-red light photoacclimation (FaRliP) in diverse cyanobacteria. Lifestyles 5:4–24

Gasper R, Schwach J, Hartmann J, Holtkamp A, Wiethaus J, Riedel N, Hofmann E, Frankenberg-Dinkel N (2017) Distinct features of cyanophage-encoded T-type phycobiliprotein lyase ΦCpeT the role of auxiliary metabolic genes. J Biol Chem 292:3089–3098

Glazer AN (1989) Light guides directional energy transfer in a photosynthetic antenna. J Biol Chem 264:1–4

González A, Angarica VE, Sancho J, Fillat MF (2014) The FurA regulon in *Anabaena* sp. PCC 7120: *in silico* prediction and experimental validation of novel target genes. Nucleic Acids Res 42:4833–4846

Gottlieb SM, Kim PW, Chang C-W, Hanke SJ, Hayer RJ, Rockwell NC, Martin SS, Lagarias JC, Larsen DS (2015) Conservation and diversity in the primary forward photodynamics of red/green cyanobacteriochromes. Biochemistry 54:1028–1042

Grossman AR, Lemaux PG, Conley PB (1986) Regulated synthesis of phycobilisome components. Photochem Photobiol 44:829–837

Grossman AR, Kehoe DM (1997) Phosphorelay control of phycobilisome biogenesis during complementary chromatic adaptation. Photosynth Res 53:95–108

Grzyb J, Latowski D, Strzalka K (2006) Lipocalins-A family portrait. J Plant Physiol 163:895–915

Gutu A, Kehoe DM (2012) Emerging perspectives on the mechanisms, regulation, and distribution of light color acclimation in cyanobacteria. Mol Plant 5:1–13

Hagiwara Y, Sugishima M, Takahashi Y, Fukuyama K (2006) Crystal structure of phycocyanobilin:ferredoxin oxidoreductase in complex with biliverdin IX alpha, a key enzyme in the biosynthesis of phycocyanobilin. Proc Natl Acad Sci USA 103:27–32

Hieber AD, Bugos RC, Yamamoto HY (2000) Plant lipocalins: violaxanthin de-epoxidase and zeaxanthin epoxidase. Biochim Biophys Acta-Protein Struct Mol Enzymol 1482:84–91

Higo A, Ikeuchi M, Ohmori M (2008) cAMP regulates respiration and oxidative stress during rehydration in *Anabaena* sp. PCC 7120. FEBS Lett 582:1883–1888

Hirose Y, Shimada T, Narikawa R, Katayama M, Ikeuchi M (2008) Cyanobacteriochrome CcaS is the green light receptor that induces the expression of phycobilisome linker protein. Proc Natl Acad Sci USA 105:9528–9533

Hirose Y, Narikawa R, Katayama M, Ikeuchi M (2010) Cyanobacteriochrome CcaS regulates phycoerythrin accumulation in *Nostoc punctiforme*, a group II chromatic adapter. Proc Natl Acad Sci USA 107:8854–8859

Hirose Y, Rockwell NC, Nishiyama K, Narikawa R, Ukaji Y, Inomata K, Lagarias JC, Ikeuchi M (2013) Green/red cyanobacteriochromes regulate complementary chromatic acclimation via a protochromic photocycle. Proc Natl Acad Sci USA 110:4974–4979

Ikeuchi M, Ishizuka T (2008) Cyanobacteriochromes: a new family of tetrapyrrole_binding photoreceptors in cyanobacteria. Photochem Photobiol Sci 7:187–292

Ishizuka T, Shimada T, Okajima K, Yoshihara S, Ochiai Y, Katayama M, Ikeuchi M (2006) Characterization of cyanobacteriochrome TePixJ from a thermophilic cyanobacterium *Thermosynechococcus elongatus* strain BP-1. Plant Cell Physiol 47:1251–1261

Jung LJ, Chan CF, Glazer AN (1995) Candidate genes for the phycoerythrocyanin α subunit lyase: biochemical analysis of *pecE* and *pecF* interposon mutants. J Biol Chem 270:12877–12884

Kahn K, Mazel D, Houmard J, Tandeau de Marsac N, Schaefer MR (1997) A role for *cpeYZ* in cyanobacterial phycoerythrin biosynthesis. J Bacteriol 179:998–1006

Kehoe DM, Grossman AR (1996) Similarity of a chromatic adaptation sensor to phytochrome and ethylene receptors. Science 273:1409–1412

Kehoe DM, Gutu A (2006) Responding to color: the regulation of complementary chromatic adaptation. Annu Rev Plant Biol 57:127–150

Kirilovsky D, Kerfeld CA (2012) The orange carotenoid protein in photoprotection of photosystem II in cyanobacteria. Biochim Biophys Acta Bioenerg 1817:158–166

Kuzin AP, Su M, Seetharaman J, Forouhar F, Wang D, Janjua H, Cunningham K, Ma L, Xiao R, Liu J, Baran MC, Acton TB, Rost B, Montelione GT, Tong L, Hunt JF (2007) Crystal structure of fatty acid-binding protein-like Ycf58 from *Thermosynecoccus elongatus*. Northeast structural genomics consortium target Ter13. In: pdb ID: 3BDR http://www.pdb.org/pdb/explore.do?structureId=3BDR

Landgraf FT, Forreite C, Hurtado P, Lamparter T, Hughes J (2001) Recombinant holophytochrome in *Escherichia coli*. FEBS Lett 508:459–462

Li L, Kehoe DM (2005) In vivo analysis of the roles of conserved aspartate and histidine residues within a complex response regulator. Mol Microbiol 55:1538–1552

Li L, Alvey RM, Bezy RP, Kehoe DM (2008) Inverse transcriptional activities during complementary chromatic adaptation are controlled by the response regulator RcaC binding to red and green light-responsive promoters. Mol Microbiol 68:286–297

Li Q, Huang Y, Zhang R, Cai T, Cai Y (2016) Medical application of *Spirulina platensis* derived C-phycocyanin. Evid Based Complementary Altern Med 2016:7803846

Liu S, Chen Y, Lu Y, Chen H, Li F, Qin S (2010) Biosynthesis of fluorescent cyanobacterial allophycocyanin trimer in *Escherichia coli*. Photosynth Res 105:135–142

Martiny AC, Coleman ML, Chisholm SW (2006) Phosphate acquisition genes in *Prochlorococcus* ecotypes: evidence for genome wide adaptation. Proc Natl Acad Sci USA 103:12552–12557

Miller CA (2007) Identification and characterization of enzymes involved in post-translational modifications of phycobiliproteins in the cyanobacterium Synechocystis sp. PCC 6803. University of New Orleans, thesis and dissertation, New Orleans, pp 1–126

Miller CA et al (2008) Biogenesis of phycobiliproteins III CpcM is the asparagine methyltransferase for phycobiliprotein β-subunits in cyanobacteria. J Biol Chem 283:19293–19300

Montgomery BL (2007) Sensing the light: photoreceptive systems and signal transduction in cyanobacteria. Mol Microbiol 64(1):16–27

Montgomery BL (2008) Shedding new light on the regulation of complementary chromatic adaptation. Cent Eur J Biol 3:351–358

Montgomery BL (2015) Light-dependent governance of cell shape dimensions in cyanobacteria. Front Microbiol 6:514

Montgomery BL (2016) Mechanisms and fitness implications of photomorphogenesis during chromatic acclimation in cyanobacteria. J Exp Bot 67:4079–4090

Moon Y-J, Kim SY, Jung K-H, Choi J-S, Park YM, Chung Y-H (2011) Cyanobacterial phytochrome Cph2 is a negative regulator in phototaxis toward UV-A. FEBS Lett 585:335–340

Morimoto K, Sato S, Tabata S, Nakai M (2003) A HEAT-repeats containing protein, IaiH, stabilizes the iron-sulfur cluster bound to the cyanobacterial IscA homologue, IscA2. J Biochem (Tokyo) 134:211–217

Narikawa R, Zikihara K, Okajima K, Ochiai Y, Katayama M, Shichida Y, Tokutomi S, Ikeuchi M (2006) Three putative photosensory LOV domains with distinct biochemical properties from the filamentous cyanobacterium *Anabaena* sp. PCC 7120. Photochem Photobiol 82:1627–1633

Narikawa R, Suzuki F, Yoshihara S, Higashi S-I, Watanabe M, Ikeuchi M (2011) Novel photosensory two-component system (PixA-NixB-NixC) involved in the regulation of positive and negative phototaxis of cyanobacterium *Synechocystis* sp PCC 6803. Plant Cell Physiol 52:2214–2224

Narikawa R, Fushimi K, Ni-Ni-Win IM (2015) Red-shifted red/green-type cyanobacteriochrome AM1_1870g3 from the chlorophyll d-bearing cyanobacterium *Acaryochloris marina*. Biochem Biophys Res Commun 461:390–395

Neunuebel MR, Golden JW (2008) The Anabaena sp. strain PCC 7120 gene all2874 encodes a diguanylate cyclase and is required for normal heterocyst development under high-light growth conditions. J Bacteriol 190:6829–6836

Ohki K, Fujita Y (1978) Photocontrol of phycoerythrin formation in the blue-green alga Tolypothrix tenuis growing in the dark. Plant Cell Physiol 19:7–15

Ohmori M, Terauchi K, Okamoto S, Watanabe M (2002) Regulation of cAMP-mediated photosignaling by a phytochrome in the cyanobacterium *Anabaena cylindrica*. Photochem Photobiol 75:675–679

Okajima K, Fukushima Y, Suzuki H et al (2006) Fate determination of the flavin photoreceptions in the cyanobacterial blue light receptor TePixD (Tll0078). J Mol Biol 363:10–18

Okamoto S, Kasahara M, Kamiya A, Nakahira Y, Ohmori M (2004) A phytochrome-like protein AphC triggers the cAMP signaling induced by far-red light in the cyanobacterium *Anabaena* sp. strain PCC7120. Photochem Photobiol 80:429–433

Ong LJ, Glazer AN (1988) Structural studies of phycobiliproteins in unicellular marine cyanobacteria. In: Stevens SE, Bryant DA (eds) Light-energy transduction in photosynthesis: higher plant and bacterial models. American Society of Plant Physiologists, Rockville, pp 102–121

Osborne BA, Raven JA (1986) Growth light level and photon absorption by cells of *Chlamydomonas rheinhardii*, *Dunaliella tertiolecta* (Chlorophyceae, Volvocales), *Scenedesmus obliquus* (Chlorophyceae, Chlorococcales) and *Euglena viridis* (Euglenophyceae, Euglenales). Br Phycol J 21:303–303

Overkamp KE, Gasper R, Kock K, Herrmann C, Hofmann E, Frankenberg-Dinkel N (2014) Insights into the biosynthesis and assembly of cryptophycean phycobiliproteins. J Biol Chem 289:691–707

Pattanaik B, Montgomery BL (2010) FdTonB is involved in the photoregulation of cellular morphology during complementary chromatic adaptation in *Fremyella diplosiphon*. Microbiology 156:731–741

Pattanaik B, Whitaker MJ, Montgomery BL (2011) Regulation of phycoerythrin synthesis and cellular morphology in *Fremyella diplosiphon* green mutants. Biochem Biophys Res Commun 413:182–188

Pattanaik B, Whitaker MJ, Montgomery BL (2012) Light quantity affects the regulation of cell shape in *Fremyella diplosiphon*. Front Microbiol 3:170

Postius C, Neuschaefer-Rube O, Haid V, Böger P (2001) N_2-fixation and complementary chromatic adaptation in non-heterocystous cyanobacteria from Lake Constance. FEMS Microbiol Ecol 37:117–125

Quest B, Hubschmann T, Sharda S, Tandeau de Marsac N, Gärtner W (2007) Homologous expression of a bacterial phytochrome. The cyanobacterium *Fremyella diplosiphon* incorporates biliverdin as a genuine, functional chromophore. FEBS J 274:2088–2098

Raffelberg S, Wang L, Gao S, Losi A, Gartner W, Nagel G (2013) A LOV-domain-mediated blue-light-activated adenylate (adenylyl) cyclase from the cyanobacterium *Microcoleus chthonoplastes* PCC 7420. Biochem J 455:359–365

Richter P, Krywult M, Sinha RP, Häder D-P (1999) Calcium signals from heterocysts of *Anabaena* sp. after UV irradiation. J Plant Physiol 154:137–139

Rockwell NC, Lagarias JC (2010) A brief history of phytochromes. Chem Phys Chem 11:1172–1180

Rockwell NC, Martin SS, Lagarias JC (2012) Red/green cyanobacteriochromes: sensors of color and power. Biochemistry 51:9667–9677

Rockwell NC, Martin SS, Gan F, Bryant DA, Lagarias JC (2015) NpR3784 is the prototype for a distinctive group of red/green cyanobacteriochromes using alternative Phe residues for photoproduct tuning. Photochem Photobiol Sci 14:258–269

Rümbeli R, Suter F, Wirth M, Sidler W, Zuber H (1987) γ-*N*-Methylasparagine in phycobiliproteins from the cyanobacteria *Mastigocladus laminosus* and *Calothrix*. FEBS Lett 221:1–2

Saunée NA, Williams SR, Bryant DA, Schluchter WM (2008) Biogenesis of phycobiliproteins. II. CpcS-I and CpcU comprise the heterodimeric bilin lyase that attaches phycocyanobilin to Cys-82 of β-phycocyanin and Cys-81 of allophycocyanin subunits in *Synechococcus* sp. PCC 7002. J Biol Chem 283:7513–7522

Savakis P, De Causmaecker S, Angerer V, Ruppert U, Anders K, Essen LO, Wilde A (2012) Light-induced alteration of c-di-GMP level controls motility of *Synechocystis* sp. PCC 6803. Mol Microbiol 85:239–251

Scheer H, Zhao KH (2008) Biliprotein maturation: the chromophore attachment. Mol Microbiol 68:263–276

Scheer H, Yang X, Zhao K-H (2015) Biliproteins and their applications in bioimaging. Proc Chem 14:176–185

Schluchter WM, Shen G, Alvey RM, Biswas A, Saunée NA, Williams SR, Mille CA, Bryant DA (2010) Phycobiliprotein biosynthesis in cyanobacteria: structure and function of enzymes involved in post-translational modification. In: Hallenbeck PC (ed) Recent advances in phototrophic prokaryotes. Springer, New York, pp 211–228

Seib LO, Kehoe DM (2002) A turquoise mutant genetically separates expression of genes encoding phycoerythrin and its associated linker peptides. J Bacteriol 184:962–970

Shen GZ, Saunee NA, Gallo E, Begovic Z, Schluchter WM, Bryant DA (2004) Identification of novel phycobiliprotein lyases in cyanobacteria. In: Niederman RA, Blankenship RE, Frank H, Robert B, van Grondelle R (eds) Photosynthesis 2004 light-harvesting systems workshop, Saint Adele, Quebec, Canada, pp 14–15

Shen G, Saunee NA, Williams SR, Gallo EF, Schluchter WM, Bryant DA (2006) Identification and characterization of a new class of bilin lyase: the *cpcT* gene encodes a bilin lyase responsible for attachment of phycocyanobilin to Cys-153 on the beta subunit of phycocyanin in *Synechococcus* sp. PCC 7002. J Biol Chem 281:17768–17778

Shen G, Leonard HS, Schluchter WM, Bryant DA (2008a) CpcM post-translationally methylates asparagine-71/72 of phycobiliprotein beta subunits in *Synechococcus* sp. PCC 7002 and *Synechocystis* sp. PCC 6803. J Bacteriol 190:4808–4817

Shen GZ, Schluchter WM, Bryant DA (2008b) Biogenesis of phycobiliproteins I. CpcS-I and CpcU mutants of the cyanobacterium *Synechococcus* sp. PCC 7002 define a heterodimeric phycocyanobilin lyase specific for β-phycocyanin and allophycocyanin subunits. J Biol Chem 283:7503–7512

Shukla A, Biswas A, Blot N, Partensky F, Karty JA, Hammad LA, Garczarek L, Gutu A, Schluchter WM, Kehoe DM (2012) Phycoerythrin-specific bilin lyase isomerase controls blue-green chromatic acclimation in marine *Synechococcus*. Proc Natl Acad Sci USA 109:20136–20141

Sidler WA (1994) Phycobilisome and phycobiliprotein structures. In: Bryant DA (ed) The molecular biology of cyanobacteria. Kluwer Academic Publication, Dordrecht, pp 139–216

Singh SP, Montgomery BL (2011) Determining cell shape: adaptive regulation of cyanobacterial cellular differentiation and morphology. Trends Microbiol 19(6):278–285

Singh SP, Montgomery BL (2012) Reactive oxygen species are involved in the morphology-determining mechanism of *Fremyella diplosiphon* cells during complementary chromatic adaptation. Microbiology 158:2235–2245

Singh SP, Montgomery BL (2014) Morphogenes *bolA* and *mreB* mediate the photoregulation of cellular morphology during complementary chromatic acclimation in *Fremyella diplosiphon*. Mol Microbiol 93:167–182

Singh SP, Montgomery BL (2015) Regulation of BolA abundance mediates morphogenesis in *Fremyella diplosiphon*. Front Microbiol 6:1215

Singh NK, Sonani RR, Rastogi RP, Madamwar D (2015) The phycobilisomes: an early requisite for efficient photosynthesis in cyanobacteria. EXCLI J 14:268–289

Song J-Y, Cho HS, Cho J-I, Jeon J-S, Lagarias JC, Park Y-I (2011) Near-UV cyanobacteriochrome signaling system elicits negative phototaxis in the cyanobacterium *Synechocystis* sp. PCC 6803. Proc Natl Acad Sci USA 108:10780–10785

Stadnichuk IN, Krasilnikov PM, Zlenko DV (2015) Cyanobacterial phycobilisomes and phycobiliproteins. Microbiology 84:101–111

Storf M, Parbel A, Meyer M, Strohmann B, Scheer H, Deng MG, Zheng M, Zhou M, Zhao KH (2001) Chromophore attachment to biliproteins: specificity of PecE/PecF, a lyase isomerase for the photoactive 3-cys-alpha 84-phycoviolobilin chromophore of phycoerythrocyanin. Biochemistry 40:12444–12456

Sugishima M, Hagiwara Y, Zhang X, Yoshida T, Migita CT, Fukuyama K (2005) Crystal structure of dimeric heme oxygenase-2 from *Synechocystis* sp. PCC6803 in complex with heme. Biochemistry 44:4257–4266

Swanson R, Glazer A (1990) Phycobiliprotein methylation: effect of the γ-*N*-methylasparagine residue on energy transfer in phycocyanin and the phycobilisome. J Mol Biol 214:787–796

Swanson RV, Zhou J, Leary JA, Williams T, de Lorimier R, Bryant DA, Glazer AN (1992) Characterization of phycocyanin produced by *cpcE* and *cpcF* mutants and identification of an intergenic suppressor of the defect in bilin attachment. J Biol Chem 267:16146–16154

Takano H, Gusella J (2002) The predominantly HEAT-like motif structure of huntingtin and its association and coincident nuclear entry with dorsal, an NF-kB/Rel/dorsal family transcription factor. BMC Neurosci 3:15

Tandeau de Marsac N (1977) Occurrence and nature of chromatic adaptation in cyanobacteria. J Bacteriol 130:82–91

Terauchi K, Ohmori M (2004) Blue light stimulates cyanobacterial motility via a cAMP signal transduction system. Mol Microbiol 52:303–309

Tooley AJ, Glazer AN (2002) Biosynthesis of the cyanobacterial light-harvesting polypeptide phycoerythrocyanin holo-α-subunit in a heterologous host. J Bacteriol 184(17):4666–4671

Tooley AJ, Cai YA, Glazer AN (2001) Biosynthesis of a fluorescent cyanobacterial C-phycocyanin holo-α subunit in a heterologous host. Proc Natl Acad Sci USA 98:10560–10565

Torrecilla I, Legané SF, Bonilla I, Fernández-Piñas F (2004) Light-to dark transitions trigger a transient increase in intracellular Ca^{2+} modulated by the redox state of the photosynthetic electron transport chain in the cyanobacterium *Anabaena* sp. PCC7120. Plant Cell Environ 27:810–819

Ulijasz AT, Cornilescu G, von Stetten D, Kaminski S, Mroginski MA, Zhang J, Bhaya D, Hildebrandt P, Vierstra RD (2008) Characterization of two thermostable cyanobacterial phytochromes reveals global movements in the chromophore-binding domain during photoconversion. J Biol Chem 283:21251–21266

Unno M, Ishikawa-Suto K, Kusaka K, Tamada T, Hagiwara Y, Sugishima M, Wada K, Yamada T, Tomoyori K, Hosoya T, Tanaka I, Niimura N, Kuroki R, Inaka K, Ishihara M, Fukuyama K (2015) Insights into the proton transfer mechanism of a bilin reductase PcyA following neutron crystallography. J Am Chem Soc 137:5452–5460

Walters KJ, Whitaker MJ, Singh SP, Montgomery BL (2013) Light intensity and reactive oxygen species are centrally involved in photoregulatory responses during complementary chromatic adaptation in *Fremyella diplosiphon*. Commun Integr Biol 6:e25005

Wilbanks SM, Wedemayer G, Glazer AN (1989) Posttranslational modifications of the β-subunit of a cryptomonad phycoerythrin. J Biol Chem 264:17860–17867

Wilde A, Churin Y, Schubert H, Borner T (1997) Disruption of a *Synechocystis* sp. PCC 6803 gene with partial similarity to phytochrome genes alters growth under changing light qualities. FEBS Lett 406:89–92

Wilks A (2002) Heme oxygenase: evolution, structure and mechanism. Antioxid Redox Signal 4:603–614

Wiltbank LB, Kehoe DM (2016) Two cyanobacterial photoreceptors regulate photosynthetic light harvesting by sensing teal, green, yellow, and red light. MBio 7(1):e02130–e02115

Yerrapragada S, Shukla A, Hallsworth-Pepin K et al (2015) Extreme sensory complexity encoded in the 10-Megabase draft genome sequence of the chromatically acclimating cyanobacterium *Tolypothrix* sp PCC 7601. Genome Announc 3:e00355–e00315

Yoshihara S, Shimada T, Matsuoka D, Zikihara K, Kohchi T, Tokutomi S (2006) Reconstitution of blue-green reversible photoconversion of a cyanobacterial photoreceptor, PixJ1, in phycocyanobilin-producing *Escherichia coli*. Biochemistry 45:3775–3784

Zhao KH, Deng MG, Zheng M, Zhou M, Parbel A, Storf M, Meyer M, Strohmann B, Scheer H (2000) Novel activity of a phycobiliprotein lyase: both the attachment of phycocyanobilin and the isomerization to phycobiliviolin are catalyzed by the proteins PecE and PecF encoded by the phycoerythrocyanin operon. FEBS Lett 469:9–13

Zhao KH, Wu D, Wang L, Zhou M, Storf M, Bubenzer C, Strohmann B, Scheer H (2002) Characterization of phycoviolobilin phycoerythrocyanin-alpha 84-cysteinlyase-(isomerizing) from *Mastigocladus laminosus*. Eur J Biochem 269:45420–44550

Zhao KH, Su P, Bohm S, Song B, Zhou M, Bubenzer C, Scheer H (2005) Reconstitution of phycobilisome core-membrane linker, LCM, by autocatalytic chromophore binding to ApcE. Biochim Biophys Acta 1706(1–2):81–87

Zhao KH, Su P, Li J, Tu JM, Zhou M, Bubenzer C, Scheer H (2006) Chromophore attachment to phycobiliprotein beta-subunits: phycocyanobilin:cysteine-beta84 phycobiliprotein lyase activity of CpeS-like protein from Anabaena Sp. PCC7120. J Biol Chem 281(13):8573–8581

Zhao KH, Su P, Tu JM, Wang X, Liu H, Ploscher M, Eichacker L, Yang B, Zhou M, Scheer H (2007a) Phycobilin: cystein-84 biliprotein lyase, a near-universal lyase for cysteine-84-binding sites in cyanobacterial phycobiliproteins. Proc Natl Acad Sci USA 104:14300–14305

Zhao KH, Zhang J, Tu JM, Böhm S, Plöscher M, Eichacker L, Bubenzer C, Scheer H, Wang X, Zhou M (2007b) Lyase activities of CpcS- and CpcT-like proteins from *Nostoc* PCC 7120 and sequential reconstitution of binding sites of phycoerythrocyanin and phycocyanin subunits. J Biol Chem 282:34093–34103

Zhao C, Gan F, Shen G, Bryant DA (2015) RfpA, RfpB, and RfpC are the master control elements of Far-Red Light Photoacclimation (FaRLiP). Front Microbiol 6:1303

Zhou J, Gasparich GE, Stirewalt VL, de Lorimier R, Bryant DA (1992) The *cpcE* and *cpcF* genes of *Synechococcus* sp. PCC 7002: construction and phenotypic characterization of interposon mutants. J Biol Chem 267:16138–16145

Stress Response of Phycobiliproteins 5

5.1 Introduction

Phycobiliproteins (PBPs) are water-soluble, brilliantly colored pigmented protein complexes functioning as predominant accessory light-harvesting complexes to harvest the photonic energy of sunlight for photosynthesis in chlorophyll *a* (Sun et al. 2003). The PBPs are exclusively found in cyanobacteria (blue-green algae, prokaryotic), rhodophytes (red algae, eukaryotes), and cryptomonads (Glazer 1994). They are autofluorescent in nature and bilin proteins covalently attached to many tetrapyrrole chromophores integrated with cysteine residues (Johnson et al. 2014). PBPs are not only the antenna complexes for light harvesting but can also be used as storage materials for carbon and nitrogen (Bryant 1987). These PBPs are composed of two types of proteins: colored phycobiliproteins (PBPs), which absorb and transmit light energy to the photosynthetic reaction centers, and nonpigmented linker proteins, which are organized into the PBPs and modulate their light absorption property (Glazer 1989). The color of the PBPs is originated mainly from the covalently bound prosthetic group. These prosthetic groups are open-chain tetrapyrrole chromophores bearing A, B, C, and D ring named as phycobilins. They are in blue-colored phycocyanobilin (PCB), purple-colored phycobiliviolin (PVB), red-colored phycoerythrobilin (PEB), and yellow-colored phycourobilin (PUB). Cyanobacteria possess four main types of PBPs: (1) allophycocyanin (APC, bluish-green color), (2) phycocyanin (PC, deep blue), (3) phycoerythrin (PE, deep red), and (4) phycocyanobilin (PCB, orange) having λmax of 650–655, 615–640, 565–575, and 575 nm, respectively (Bryant et al. 1979). According to evolution of PBPs, specific variability of PBPs is due to nucleotide variation at third position of GC codon (Kannaujiya et al. 2014a). In the past few years, PBPs contribute a major role in the development of commercial colored food products, nutraceuticals, and biotechnological products. They are widely used in pharmaceutical, fluorescent, and cosmetic industry for utilization of brilliant color property of PBPs in various commercial and scientific products. Recently, it has been approved for antioxidant, anti-inflammatory, neuroprotective, and hepatoprotective properties for clinical

V.K. Kannaujiya et al., *Phycobiliproteins: Recent Developments and Future Applications*, https://doi.org/10.1007/978-981-10-6460-9_5

application (Spolaore et al. 2006). Cyanobacteria have the ability to regulate the composition and property of PBPs in response to various abiotic environmental signals like nutrient availability, light intensity, and temperature (Prassana et al. 2004). Abiotic stress plays an important role in sustainable production of PBPs from cyanobacteria. In this chapter, we describe certain abiotic stress that has effects on composition of PBPs in cyanobacteria.

5.2 Abiotic Stress

5.2.1 Light Stress

Cyanobacteria are pioneer prokaryotic organisms that play an important role as primary producers in both aquatic and terrestrial ecosystems (Häder et al. 2011). The nitrogen-fixing cyanobacteria play an indispensable role in acclimation of carbon and nitrogen economy (C/N ratio) in paddy fields (Singh et al. 2013). Solar radiation consists mainly of infrared (700–3000 nm), photosynthetically active radiation (PAR; 400–700 nm), and ultraviolet (UV; 100–400 nm) radiation (Fig. 5.1). Based on wavelength characteristics, ultraviolet radiation has been divided into ultraviolet A (UVA; 315–400), ultraviolet B (UVB; 280–315 nm), and ultraviolet C (UVC; 100–280 nm) in solar spectrum. A very small magnitude of solar ultraviolet radiation contributes UVC (0%), UVB (<1%), and UVA (<7%) in total irradiance impinging on the Earth's surface. Highly energetic UVC does not reach the Earth's surface due to complete absorption by atmospheric ozone layer. UV protective ozone layer of our green planet (Earth) is depleting due to unusual human activities by release of pollutants that leads to the depletion of stratospheric ozone layer and consequent increase in harmful ultraviolet B radiation (Kerr and McElory 1993; Solomon 1999). Apart from that, unpolluted terrestrial, aquatic ecosystems and anthropogenic sources naturally produce substantial amount of reactive nitrogen species (RNS) such as nitric oxide (NO) peroxynitrite (ONOO-) and nitrous oxide (N_2O) which further contribute to damage of the ozone layer (Kramlich and Linak 1994). UV radiation has high intensity of energy that is able to produce many types of radicals such as superoxide anion radicals, hydrogen peroxide, hydroperoxyl radicals, hydroxyl radicals, and hypochlorous radicals that cause several damage in biomolecules of proteins, lipids, and nucleic acid of proteins (Halliwell and Gutteridge 2007; Singh et al. 2014; Kannaujiya et al. 2014b).

The continuous increase in UVB radiation causes alterations in biologically important proteins, nucleic acids, and other relevant molecules, thereby affecting a number of vital physiological and biochemical functions in living organisms (Sinha et al. 2002). However, in a tropical ecosystem, no significant changes in ozone density are observed during period of 2008–2012 (Bais et al. 2015).

Extensive literature survey shows that UVB radiation causes various damaging effects on cyanobacterial cells and its photosynthetic apparatus (Sinha et al. 1995; He and Häder 2002; Unsal-Kacmaz et al. 2002). Gao et al. (2007) recorded about 40% growth inhibition in *Anabaena* sp. PCC7120 and certain other cyanobacteria

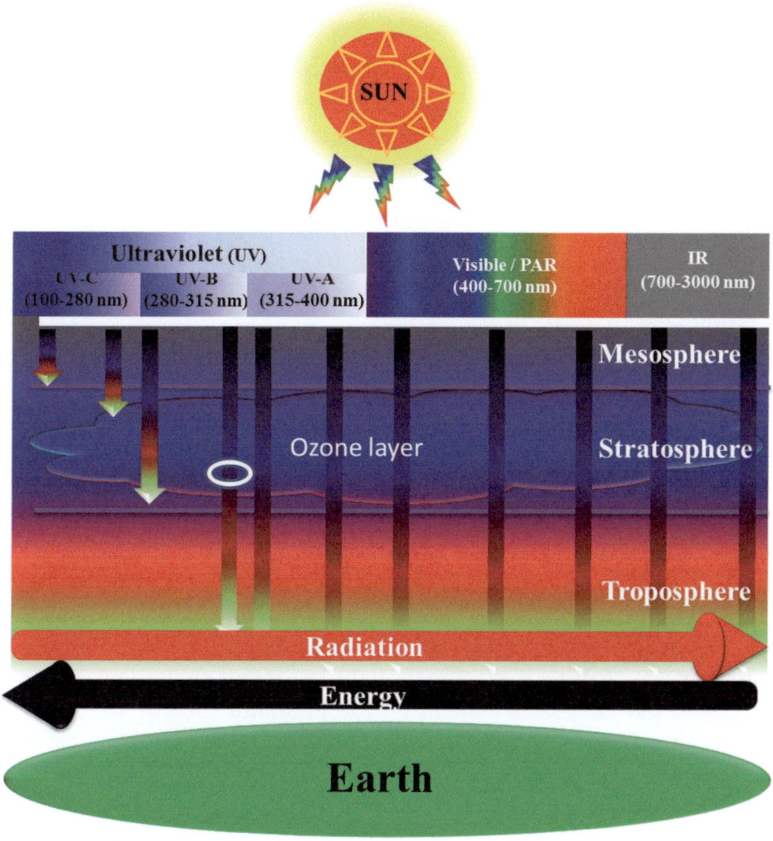

Fig. 5.1 Electromagnetic spectrum of sun showing various radiation wavebands

under solar UV irradiation. UVB radiation also induces the production of ROS as detected by using the ROS-sensitive probe 2′,7′-dichlorodihydrofluorescein diacetate (DCFH-DA) (Rastogi et al. 2010). UVB induces generation of ROS by photodynamic action and inhibition of the electron by damaging receptor site or enzymes associated with the electron transport chain during photosynthesis (He and Häder 2002). The inhibition of growth and killing of cyanobacteria by UVB radiation has been reported by Döhler (1986). It has been demonstrated that sublethal exposure level of UVB produces intracellular damage that affects the growth and the endogenous rhythms of microorganisms. Any damage in the photosynthetic pigments severely affects the photosynthesis processes. PE-rich strain of *Nostoc* sp. was found to be more tolerant under UVB irradiation as compared to PC-rich strain (Tyagi et al. 1992). Döhler (1986) has previously reported bleaching of photosynthetic pigments by UVB. After being irradiated with UVB radiation, the photosynthetic activity (Fv/Fm), cellular total carbohydrate, EPS (exopolysaccharide), and sucrose production have been shown to be decreased, while reducing sugar, ROS, and malondialdehyde (MDA) production and DNA strand break increased significantly (Chen et al. 2008). This radiation-induced genetic changes result in DNA damage and change in protein contents (Zhou et al. 2009). It is well established that UVB irradiation has pronounced effects on freshwater cyanobacteria (Sinha et al. 1995; Rajagopal et al. 1998; Rinalducci et al. 2006; Six et al. 2007). It has been reported that UV irradiation affects the anchor linker polypeptides of the PBPs (Sinha et al. 1995; Sah et al. 1998; Kannaujiya and Sinha 2015). UV irradiation also affects the integrity of PBPs and pigment in marine picocyanobacterial species such as *Synechococcus* sp. WH8102 (Six et al. 2007). Kulandaivelu et al. (1989) have demonstrated that UVB exposure inhibits the oxygen evolution and also disturbs spectral properties of PBPs in *Synechococcus* sp. by partial uncoupling of energy transfer. High intensity of UVB irradiation also accelerated the damage of linker or anchor polypeptides in *Synechococcus* sp. PCC 7942 (Pandey et al. 1997). Sah et al. (1998) reported that low dose of UVB irradiation altered the anchor polypeptide (75 kDa) between protein and pigment in *Synechococcus* sp. 7942. Similarly, Rajagopal et al. (1998) reported that anchor polypeptide was altered in PBPs of *Spirulina platensis* after UVB irradiation. The fluorescent nature of PC/PE is dependent on cysteine-linked integration of linear tetrapyrrole chromophores in α and β monomers. The high intensity of PAR and UV radiation reduces the fluorescent nature of PBPs (Rastogi et al. 2015; Kannaujiya and Sinha 2017a). Apart from UV radiation, the intensity of light also affects PBP composition in cyanobacteria.

The divergence of growth at different light intensities shows that 25 μmol photons/m^2/s was the best-suited intensity for optimum growth of *Spirulina* sp. (Tomasseli et al. 1995, 1997;), *Synechococcus* NKBG 042902 (Takano et al. 1995), and *Synechocystis* (Hong and Lee 2008). However, optimum growth was recorded after 50% reduction in light intensity (12.5 μmol photons/m^2/s) in cyanobacterium *Nostoc* UAM 206 (Poza-Carrion et al. 2001) and *Nostoc muscorum* (Ranjitha and Kaushik 2005). Interestingly, it has been suggested that bilin proteins for PBPs are more stimulated in low light intensities due to minimum energy consumption in maintenance of the photosystem (Grossman et al. 1993). Eukaryotic red algae required higher irradiance for growth and development such as 40 μmol photons/m^2/s intensity optimum for *Gracilaria tenuistipitata* (Carnicas et al. 1999) and 65 μmol photons/m^2/s for *Audouinella*, *Batrachospermum*, and *Compsopogon* (Zucchi and Neechi 2001). Certain cyanobacteria such as *Arthronema africanum* required high light intensity up to 150 μmol photons/m^2/s for optimum production of PBPs (Chaneva et al. 2007).

The optimum intensity of color of light may enhance the productivity of PBPs in certain cyanobacteria. It has been reported red light may affect the growth of cyanobacteria and stimulate production of PC in *Anacystis nidulans* (Lonneborg et al. 1985), *Calothrix* 7601 (Liotenberg et al. 1996), *Nostoc* UAM206 (Poza-Carrion et al. 2001), *Nostoc muscorum* (Ranjitha and Kaushik 2005), and *Synechococcus* (Takano et al. 1995). The wavelength of blue spectrum has shown stimulatory effects for PE synthesis in red algae such as *Rhodella reticulata* (Mihova et al. 1996), *Porphyra leucosticta* (Tsekos et al. 2002), *Chondrus crispus* (Franklin et al.

2002), and *Halymenia floresii* (Godinez-Ortega et al. 2008). The dynamic fluctuation or rhythmic alteration of light and dark period of light induces to change accessory light-harvesting capacity for photosynthesis (Kono and Terashima 2014). The fluctuation in UVB radiation suggests induction of photoprotective mechanism to protect PBPs from damage (Chukhutsina et al. 2015; Kannaujiya and Sinha 2017a, b).

5.2.2 Temperature Stress

Temperature is the most fundamental physiological factor that affects physiological, biochemical, and metabolic processes of an organism. The survivability, growth, and stress tolerance of an organism dependent on temperature usually vary from species to species. However, sudden fluctuation in temperature may exert negative effects on growth and development of organism. In aquatic system, high temperature reduces availability of free oxygen which may be a primary cause of stress in photosynthesis (Brock 1978). Hemlata and Fatma (2009) have found optimum temperature about 30 °C for synthesis of PBPs in cyanobacterium *Anabaena* NCCU-9. However, alteration in optimum temperature (30 °C) may reduce the growth significantly with optimum yield of PBPs. Moreover, other scientists reported the optimum temperature to be 36 and 37 °C for *Arthronema africanum* (Chaneva et al. 2007) and *Synechococcus* (Sakamoto and Bryant 1998), respectively. Upon exposure to higher temperature, the cyanobacterium *Synechococcus* sp. strain PCC 7942 will produce HspA small heat shock proteins that interact with PC of PBPs and suppress the inactivation of functional properties by heat-enabled denaturation (Nakamoto and Honma 2006). They also thrive well at high temperature (Richa and Sinha 2015).

5.2.3 Biochemical Stress

5.2.3.1 Nitrogen Stress

Cyanobacteria have special requirements regarding nitrogen sources, such as NO_3^-, NH_4 or carbamide, and $NaNO_3$. *Anabaena* NCCU-9 produced the highest amount of PBPs under nitrogen-free environment (Hemlata and Fatma 2009). The cyanobacterium *Fischerella* sp. produced more amounts of PBPs under nitrate or ammonium medium as compared to nitrogen-free growth media (Soltani et al. 2007). However, in another cyanobacterium like *Anabaena* sp. PCC 7120, the amount of PBPs merely exceeded in nitrogen-free media as compared to nitrate medium (Loreto et al. 2003). Hemlata and Fatma (2009) have reported about toxicity of ammonium in growth medium, which reduces growth that leads to cell death in *Spirulina platensis*, *Nostoc* sp., *Calothrix* sp. PCC7601, and *Nodularia spumigena* (Belkin and Boussiba 1991; Liotenberg et al. 1996).

5.2.3.2 Salt Stress

The maintenance of salt concentration is crucial for proper cell functioning, ion regulation, membrane potential, osmotic balance, and metabolic activity. Salinity adversely affects 19.5% of the irrigated agricultural lands worldwide and shows serious threat to crop productivity (Pandhal et al. 2008). Salt-induced effects have extensively been studied in certain salt-tolerant cyanobacteria such as *Synechocystis* PCC 6803 and *Euhalothece* sp. (Jeanjean et al. 1993; Huang et al. 2006; Pandhal et al. 2009). The composition and function of PBPs in cyanobacteria changed in response to stress conditions (Grossman et al. 1993). Salt stress mainly decreases in PC concentration, and thereby energy transfer between PBPs and PSII may be interrupted (Schubert and Hagemann 1990; Schubert et al. 1993; Lu et al. 1999; Lu and Vonshak 2002). It was found that the lowest concentration of salt (10 mM) resulted to an increase in PBP content in cyanobacterium *Anabaena* NCCU-9 as compared to untreated sample, while further increase in salt concentration resulted to gradual decline in growth (Hemlata and Fatma 2009). To increase salt concentration, the ionic movement of sodium ions might be increased, and induced detachment of PBPs leads to inhibition of energy transfer reaction between PBPs and PSII (Verma and Mohanty 2000; Rafiqul et al. 2003).

5.2.3.3 Pesticide Stress

The toxicity of pesticides is a major concern on growth and development of cyanobacteria including PBPs. Pesticides like malathion have shown bleaching effects on PBPs; therefore concentration becomes rapidly declined, while other pesticides like chlorpyrifos slow down growth of cyanobacteria (Hemlata and Fatma 2009). Certain other pesticides also show an inhibitory response on PBPs of cyanobacteria such as effect of endosulfan on *Plectonema boryanum* (Prasad et al. 2005) arozin, alachlor, and butachlor on *Anabaena variabilis* (Singh and Dutta 2005); thiobencarb on *Nostoc sphaeroides* (Xia 2005); cypermethrin on *Anabaena doliolum* (Mohapatra et al. 2003); lindane on *Anabaena* sp. (Babu et al. 2001); thiobencarb on *Anabaena variabilis* (Battah et al. 2001) organophosphate on *Synechocystis* PCC6803 (Mohapatra and Scheiwer 2000) and endosulfan on *Nostoc linckia* (Satish and Tiwari 2000). The pesticide-induced reduction of photosynthetic activity may be due to inhibition or bleaching effects on the photosystem by excessive production of ROS.

5.2.3.4 Nutrient Stress

Cyanobacteria show a growth inhibitory response by enrichment and limitation in optimum nutrient concentration. Sulfur is a limiting element for synthesis and functionalization of PBPs. A poor source of sulfur nutrient or thiolated amino acids may induce degradation in PBP content of cyanobacteria (Schwarz and Grossman 1998). Sulfur-deprived medium for culture of *Synechocystis* PCC 6803 shows rapid reduction in PBP concentration as compared to phosphorus-deprived culture of same cyanobacteria (Richaud et al. 2001). Similarly, limitation of any optimum nutrient concentration may inhibit activity of photosynthetic membranes (Bhaya et al. 2000). In sulfur-deprived medium *Synechococcus* sp. has induced expression of gene called *nblA* having proteolysis activity which plays a crucial role in the degradation of PBPs (Collier and Grossman 1994). There are dramatic changes in color of cyanobacteria from bluish green to yellowish green under deprivation of nutrient supplements (Allen and Smith 1969).

5.2.4 Metal Stress

Pollution has great impact on physical, chemical, and biological basis of environment which leads to global warming, water pollution, and air pollution and a much more inhibitory response on the life of an organism. It has deleterious effects on plants, microorganism consortium, and humans to a great extent. Heavy metals play a leading role to induce toxicity of undesirable pollutants on an organism. The toxicity of heavy metals may further increase due to their nonbiodegradable nature and ability to enter into the food chain (Nellesson and Fletcher 1993; Salt and Rauser 1995). Heavy metals severely affect normal physiological processes by interference of essential enzyme activity and pigment-protein developmental process (Bertrand and Guary 2002). The major effect includes changes in the pigment composition which affects net photosynthesis rate by inhibition of carboxylation mechanism that ultimately inhibits photosystem II (PS II) activity. However, a few heavy metals as micronutrients critically play a significant role in the growth of cyanobacteria. Micronutrient heavy metals include boron, manganese, zinc, molybdenum, copper, and cobalt required in very low quantities. Boron is a rare metal and keenly required by plants, but has no role in the growth of fungi as well as animals. Anderson and Jordan (Anderson and Jordan 1961) reported that boron is essential for the dinitrogen fixation in *Azotobacter*. Boron is an essential micronutrient for rapid growth and development in nitrogen-deprived condition in certain cyanobacteria like *Nostoc muscorum*, *Calothrix parietina*, and *Anabaena cylindrica* (Gerloff 1968). The boron-induced growth of cyanobacteria may signify rapid synthesis of PBPs. Manganese and zinc are essential elements for the growth of cyanobacteria particularly in different enzymatic reactions involved in photosynthetic processes (Casarett and Doull 1980). Their deficiency may cause inhibition of chlorophyll biosynthesis (Csatorday et al. 1984). Manganese is an important constituent of water-splitting system that provides electrons to PSII as well as cofactors for different enzymes. Zinc is an essential micronutrient that is involved in numerous physiological processes in the enzymes oxidoreductases, hydrolases, lyases, and ligases (Barak and Helmke 1999). It also plays a critical role in the function of metalloenzymes (Chaney 1993). Zinc deficiency affects mitotic division of cells which results into cells being large and aberrant in appearance. As an industrial effluent zinc facilitate changes in structural integrity of photosynthetic activity of cells. Metal-induced changes in the arrangement and structure of the light-harvesting complex in cyanobacteria have been well documented (Sersen and Kralova 2001). Murthy and Mohanty (Murthy and Mohanty 1991) have found that energy transfer from PC to chlorophyll *a* (Chl *a*) is rapidly inhibited by heavy metals. This eventually could affect the synthesis of PC and carotenoids as reported by Atri and Rai (2003) after studying the effects of cadmium on three different cyanobacterial genera. Hemlata and Fatma (2009) have reported decline in PBP content after treatment with ions. There was maximum decline in PBP amount found in Cr^{+6} and subsequently Cd^{+2}, Pb^{+2}, Ni^{+2}, Cu^{+2}, and Zn^{+2} ions. However, PBP content was significantly increased up to four times by the treatment with Pb^{+2} stress in *Microchaete tenera* (Zaccaro et al. 2000).

5.3 Conclusion

The diversity of cyanobacteria is very broad and widely distributed in various ecological habitats. Stress tolerance behavior of PBPs is varying from species to species. In laboratory condition, an optimum level of temperature, pH, color of wavelength, and basic nutrient medium is used for growth of many cyanobacteria. The changes in light spectrum like UV radiation have several effects on PBPs of many cyanobacterial communities residing in different habitats. Cyanobacteria existing in harsh conditions seem to be less prone to various physiological stresses. Most of the metals exhibit toxicity on PBPs and inhibited growth of cyanobacteria. Concurrent abiotic stress shows rapid degradation in PBP composition, which severely affects biotechnological production for commercial utilization in industry. Thus, close adjustment of abiotic stress of the environment may enhance the productivity of PBPs from any cyanobacterial species.

References

Allen MM, Smith AJ (1969) Nitrogen chlorosis in blue green algae. Arch Mikrobiol 69:114–120

Anderson GR, Jordan JV (1961) Boron: a non-essential growth factor for *Azotobacter chroococcum*. Soil Sci 92:113–116

Atri N, Rai LC (2003) Differential responses of three cyanobacteria to UV-B. Cd J Microbiol Biotech 13:544–551

Babu GS, Hans RK, Singh J, Viswanathan PN, Joshi PC (2001) Effect of lindane on the growth and metabolic activities of cyanobacteria. Ecotoxicol Environ Saf 48:219–221

Bais AF, McKenzie RL, Bernhard G, Aucamp PJ, Ilyas M, Madronichf S, Tourpali K (2015) Ozone depletion and climate change: impacts on UV radiation. Photochem Photobiol Sci 14:19–52

Barak P, Helmke PA (1999) The chemistry of Zn. In: Robson AD (ed) Zn in soils and plants. Developments in plants and soils sciences. Kluwer Academic Publishers, Dordrecht, pp 1–13

Battah MG, Shabana EF, Kobbia JA, Eldel HM (2001) Differential effects of thiobencarb toxicity on the growth and photosynthesis of *Anabaena variabilis* with changes in phosphate level. Ecotoxicol Environ Saf 49:235–239

Belkin S, Boussiba S (1991) High internal pH conveys ammonia resistance in Spirulina Platensis. Bioresour Technol 38:167–169

Bertrand M, Guary JC (2002) How plants adopt their physiology to an excess of metals. In: Pessarakli M (ed) Handbook of plant and crop physiology, 2nd edn. Marcel Dekker, New York, pp 751–761

Bhaya D, Schwarz R, Grossman AR (2000) Molecular responses to environmental stresses. In: Whitton BA, Potts M (eds) The ecology of cyanobacteria. Kluwer, Dordrecht, pp 397–442

Brock TD (1978) Thermophilic microorganisms and life at high temperatures. Springer, New York

Bryant DA (1987) The cyanobacterial photosynthetic apparatus: comparisons to those of higher plants and photosynthetic bacteria. In: Platt T, Li WKW (eds) Photosynthetic picoplankton. Department of Fisheries and Oceans, Ottawa, pp 423–500

Bryant DA, Guglielmi G, Tandeau de Marsac N, Castlets AM, Cohen-Bazire G (1979) The structure of cyanobacterial phycobilisomes: a model. Arch Microbiol 123:113–127

Carnicas E, Jimenez C, Niell FX (1999) Effects of changes of irradiances on the pigment composition of *Gracilaria tenuistipitata* var. Liui Zhang et Xia. J Photochem Photobiol B 50:149–158

Casarett L, Doull J (1980) Toxicology, 2nd edn. Macmillan Publishing Co, New York, pp 401–611

Chaneva G, Furnadzhieva S, Minkova K, Lukavsky J (2007) Effect of light and temperature on the cyanobacterium *Arthronema africanum* a prospective phycobiliprotein producing strain. J Appl Phycol 19:537–544

Chaney RL (1993) Zn is soils and plants. In: Robson AD (ed) Developments in plant and soil sciences. Springer, Netherlands, pp 45–57

Chen LZ, Wang GH, Hong S, Liu A, Li C, Liu YD (2008) UV-B induced oxidative damage and protective role of exopolysaccharide in desert cyanobacterium *Microcoleus vaginatus*. J Inter Plant 1:7444–7909

Chukhutsina V, Bersanini L, Aro EM, van Amerongen H (2015) Cyanobacterial light-harvesting phycobilisomes uncouple from photosystem I during dark-to-light transitions. Sci Rep 5:14193

Collier JL, Grossman AR (1994) A small peptide triggers complete degradation of light harvesting phycobiliproteins in nutrients deprived cyanobacteria. EMBO J 13:1039–1047

Csatorday K, Gombos Z, Szalontai B (1984) Manganese and cobalt toxicity in chlorophyll biosynthesis. Proc Natl Acad Sci U S A 81:476–478

Döhler G (1986) Effects of UV-B radiation on the nitrogen metabolism of several marine diatoms. J Plant Physiol 118:391–400

Franklin LA, Krabs G, Kuhlenkamp P (2002) Blue light and UV radiation control the synthesis of mycosporine like amino acids in *Chondrus crispus* (Florideophyceae). J Phycol 37:257–270

Gao K, Yu H, Brown MT (2007) Solar PAR and UV radiation affects the physiology and morphology of the cyanobacterium *Anabaena* sp. PCC 7120. J Photochem Photobiol B Biol 89:117–124

Gerloff GC (1968) The comparative boron nutrition of several green and blue-green algae. Physiol Plant 21:369–377

Glazer AN (1989) Light guides. J Biol Chem 264:1–4

Glazer AN (1994) Adaptive variations in phycobilisome structure. Adv Mol Cell Biol 10:119–149

Godinez-Ortega JL, Snoeijs P, Robledo D, Freile-Pelegrin Y, Pedersen M (2008) Growth and pigment composition in the red algae *Halymenia floresii* cultured under different light qualities. J Appl Phycol 20:253–260

Grossman AR, Schaer M, Chiang G, Collier J (1993) Environmental effects on the light harvesting complex of cyanobacteria. J Bacteriol 175:575–582

Häder D-P, Helbling EW, Williamson CE, Worrest RC (2011) Effects of UV radiation on aquatic ecosystems and interactions with climate change. Photochem Photobiol Sci 10:242–260

Halliwell B, Gutteridge J (2007) Free radicals in biology and medicine. Oxford university press, Oxford

He YY, Häder D-P (2002) Reactive oxygen species and UV-B: effect on cyanobacteria. Photochem Photobiol Sci 1:729–736

Hemlata, Fatma T (2009) Screening of cyanobacteria for phycobiliproteins and effect of different environmental stress on its yield. Bull Environ Contam Toxicol 83:509–515

Hong SJ, Lee CG (2008) Statistical optimization of culture media for production of phycobiliprotein by *Synechocystis* sp. PCC 6701. Biotechnol Bioprocess Eng 13:491–498

Huang F, Fulda S, Hagemann M, Norling B (2006) Proteomic screening of salt-stress-induced changes in plasma membranes of *Synechocystis* sp. strain PCC 6803. Proteomics 6:910–920

Jeanjean R, Matthijs HCP, Onana B, Havaux M, Joset F (1993) Exposure of the cyanobacterium *Synechocystis* PCC6803 to salt stress induces concerted changes in respiration and photosynthesis. Plant Cell Physiol 34:1073–1079

Johnson EM, Kumar K, Das D (2014) Physicochemical parameters optimization, and purification of phycobiliproteins from the isolated *Nostoc* sp. Bioresour Technol 166:541–547

Kannaujiya VK, Sinha RP (2015) Impacts of varying light regimes on phycobiliproteins of *Nostoc* sp. HKAR-2 and *Nostoc* sp. HKAR-11 isolated from diverse habitats. Protoplasma 252:1551–1561

Kannaujiya VK, Sinha RP (2017a) Detection of free thiols and fluorescence response of phycoerythrin chromophore after ultraviolet-B radiation stress. J Fluoresc 27:561–567

Kannaujiya VK, Sinha RP (2017b) Impacts of diurnal variation of ultraviolet-B and photosynthetically active radiation on phycobiliproteins of the hot-spring cyanobacterium *Nostoc* sp. strain HKAR-2. Protoplasma 254:423–433

Kannaujiya VK, Rastogi RP, Sinha RP (2014a) GC constituents and relative codon expressed amino acid composition in cyanobacterial phycobiliproteins. Gene 546:162–171

Kannaujiya VK, Richa, Sinha RP (2014b) Peroxide scavenging potential of ultraviolet-B-absorbing mycosporine-like amino acids isolated from a marine red alga *Bryocladia* sp. Front Environ Sci 2:1–8

Kerr JB, McElory CT (1993) Evidence for large upwards trends of ultraviolet-B radiation linked to ozone depletion. Science 262:1032–1034

Kono M, Terashima I (2014) Long-term and short-term responses of the photosynthetic electron transport to fluctuating light. J Photochem Photobiol B Biol 137:89–99

Kramlich JC, Linak WP (1994) Nitrous oxide behavior in the atmosphere, and in combustion and industrial systems. Prog Energy Combust Sci 20:149–202

Kulandaivelu G, Gheetha V, Periyanan S (1989) Inhibition of energy transfer reactions in cyanobacteria by different ultraviolet radiation. Singhal GS, Barber J, Dilley RA, Govindjee, Haselkorn R, Mohanty P Photosynthesis, molecular biology and bioenergetics, Spriger-Verlag, New York, 305–313

Liotenberg S, Campbell D, Rippka R, Houmard J, Tandeu de Marsac N (1996) Effect of the nitrogen source on phycobiliprotein synthesis and cell reserves in a chromatically adapting filamentous cyanobacterium. Microbiology 142:611–622

Lonneborg A, Lind AK, Kalia SR, Gustafsson P, Oquist G (1985) Acclimation process in the light-harvesting system of the cyanobacterium *Anacystis nidulans* following a light shift from white to red light. Plant Physiol 78:110–114

Loreto C, Rosales N, Bermúdez J, Morales E (2003) Pigment and protein production of the cyanobacterium *Anabaena* PCC 7120 in relation to nitrogen concentration and irradiance. Gayana Bot 60:83–89

Lu C, Vonshak A (2002) Effects of salinity on photosystem II function in cyanobacterial *Spirulina platensis* cells. Physiol Plant 114:405–413

Lu C, Torzilo G, Vonshak A (1999) Kinetic response of photosystem II photochemistry in cyanobacterium *Spirulina platensis* to high salinity is characterized by two distinct phases. Aust J Plant Physiol 26:283–292

Mihova SG, Georgiev DI, Minkova KM, Tchernov AA (1996) Phycobiliproteins in *Rhodella reticulata* and photoregulatory effects on their content. J Biotechnol 48:251–257

Mohapatra PK, Scheiwer U (2000) Dimethoate and quinalphos toxicity: pattern of photosynthetic pigment degradation and recovery in *Synechococcus* sp. PCC 6803. Algol Stud 99:79–94

Mohapatra PK, Patra S, Samantaray PK, Mohanty RC (2003) Effect of the pyrethroid insecticide cypermethrin on photosynthetic pigments of the cyanobacterium *Anabaena doliolum* Bhar. Pol J Environ Stud 12(2):207–212

Murthy SDS, Mohanty P (1991) Mercury induces alteration of energy transfer in phycobilisomes by selectively affecting the pigment protein, phycocyanin in the cyanobacterium *Spirulina platensis*. Plant Cell Physiol 32:231–237

Nakamoto H, Honma D (2006) Interaction of a small heat shock protein with light-harvesting cyanobacterial phycocyanins under stress conditions. FEBS Lett 580:3029–3034

Nellesson H, Fletcher JS (1993) Assessment of publish eds. Literature on the uptake, accumulation, and translocation of heavy metals by vascular. Plants Chemosphere 9:1669–1680

Pandey R, Chauhan S, Singhal GS (1997) UVB-induced photodamage to phycobilisomes of *Synechococcus* sp. PCC 7942. J Photochem Photobiol B Biol 40:228–232

Pandhal J, Biggs C, Wright P (2008) Proteomics with a pinch of salt: a cyanobacterial perspective. Saline Syst 4:1

Pandhal J, Ow SY, Wright PC, Biggs CA (2009) Comparative proteomics study of salt tolerance between a non-sequenced extremely halotolerant cyanobacterium and its mildly halotolerant relative using in vivo metabolic labeling and in vitro isobaric labeling. J Proteome Res 8:818–828

Poza-Carrion C, Fernadez-Valiente E, Fernadez-Pinas F, Leganes F (2001) Acclimation of photosynthetic pigments and photosynthesis of the cyanobacterium *Nostoc* sp. strain UAM 206 to combined fluctuations of irradiance, pH and inorganic carbon availability. J Plant Physiol 158:1455–1461

Prasad SM, Kumar D, Zeeshan M (2005) Growth, photosynthesis, active oxygen species and antioxidants responses of paddy field cynobacterium *Plectonema boryanum* to endosulfan stress. J Gen Appl Microbiol 51:115–123

Prassana R, Pabby A, Saxena S, Singh PK (2004) Modulation of pigment profiles of *Calothrix elenkenii* in response to environmental changes. J Plant Physiol 161:1125–1132

Rafiqul IM, Hassan A, Sulebele G, Orosco CA, Roustaian P, Jalal KCA (2003) Salt stress culture of blue green algae *Spirulina fusiformis*. Pak J Biol Sci 6:648–650

Rajagopal S, Jha IB, Murthy SDS, Mohanty P (1998) Ultraviolet-B effects on *Spirulina platensis* cells: modification of chromophore-protein interaction and energy transfer characteristics of phycobilisomes. Biochem Biophys Res Commun 249:172–177

Ranjitha K, Kaushik BD (2005) Influence of environmental factors on accessory pigments of Nostoc muscorum. Indian J Microbiol 45:67–69

Rastogi RP, Singh SP, Häder D-P, Sinha RP (2010) Detection of reactive oxygen species (ROS) by the oxidant-sensing probe 2′,7′-dichlorodihydrofluorescein diacetate in the cyanobacterium *Anabaena variabilis* PCC 7937. Biochem Biophys Res Commun 397:603–760

Rastogi RP, Sonani RR, Madamwar D (2015) Effects of PAR and UV radiation on the structural and functional integrity of phycocyanin, phycoerythrin and allophycocyanin isolated from the marine cyanobacterium *Lyngbya* sp. A09DM. Photochem Photobiol 91:837–844

Richaud C, Zabulon G, Jodder A, Thomas JC (2001) Nitrogen and sulphur starvation differentially affects phycobilisomes degradation and expression of the nblA gene in *Synechocystis* strain PCC 6803. J Bacteriol 183:2989–2994

Richa, Sinha RP (2015) Biochemical characterization of sunscreening mycosporine-like amino acids from two *Nostoc* species inhabiting diverse habitats. Protoplasma 252(1):199–208

Rinalducci S, Hideg E, Vass I, Zolla L (2006) Effect of moderate UV-B irradiation on *Synechocystis* PCC 6803 biliproteins. Biochem Biophys Res Commun 341:1105–1112

Sah JF, Krishna KB, Srivastava M, Mohanty P (1998) Effects of ultraviolet-B radiation on phycobilisomes of *Synechococcus* PCC 7942: alterations in conformation and energy transfer characteristics. Biochem Mol Biol Int 44:245–257

Sakamoto T, Bryant DA (1998) Growth at low temperature causes nitrogen limitation in the cyanobacterium *Synechococcus* sp. PCC 7002. Arch Microbiol 169:10–19

Salt DE, Rauser WE (1995) Mg-ATP dependent transport of phytochelatins across the tonoplast of oat roots. Plant Physiol 107:1293–1301

Satish N, Tiwari GL (2000) Pesticide tolerance in *Nostoc linckia* in relation to the growth and nitrogen fixation. Proc Natl Am Sci 70:319–323

Schubert H, Hagemann M (1990) Salt effects on 77K fluorescence and photosynthesis in the cyanobacterium *Synechocystis* sp. PCC 6803. FEMS Microbiol Lett 71:169–172

Schubert H, Fulda S, Hagemann M (1993) Effects of adaptation to different salt concentrations on photosynthesis and pigmentation of the cyanobacterium *Synechocystis* sp. PCC 6083. J Plant Physiol 142:291–295

Schwarz R, Grossman AR (1998) A response regulator of cyanobacteria integrates diverse environmental signals and is critical for survival under extreme conditions. Proc Nat Acad Sci USA 95:1108–1113

Sersen F, Kralova K (2001) New facts about $CdCl_2$ action on the photosynthetic apparatus of spinach chloroplast and its comparison with $HgCl_2$ action. Photosynthetica 39:575–580

Singh S, Dutta P (2005) Growth and survival potentials of immobilized diazotrophic cyanobacteria isolates exposed to common rice field herbicides. World J Microbiol Biotechnol 2:441–446
Singh G, Babele PK, Sinha RP, Tyagi MB, Kumar A (2013) Enzymatic and non-enzymatic defense mechanisms against ultraviolet-B radiation in two *Anabaena* species. Process Biochem 48:796–802
Singh SP, Rastogi RP, Häder D-P, Sinha RP (2014) Temporal dynamics of ROS biogenesis under simulated solar radiation in the cyanobacterium *Anabaena variabilis* PCC 7937. Protoplasma 251:1223–1230
Sinha RP, Lebert M, Kumar A, Kumar HD, Häder D-P (1995) Spectroscopic and biochemical analyses of UV effects of phycobilisomes of *Anabaena* sp. and *Nostoc carmium*. Bot Acta 108:87–92
Sinha RP, Richter P, Faddoul J, Braun M, Häder D-P (2002) Effects of UV and visible light on cyanobacteria at the cellular level. Photochem Photobiol Sci 1:553–559
Six C, Joubin L, Partensky F, Holtzendorff J, Garczarek L (2007) UV-induced phycobilisome dismantling in the marine picocyanobacterium *Synechococcus* sp. WH8102. Photosynth Res 92:75–86
Solomon S (1999) Stratospheric ozone depletion: a review of concepts and history. Rev Geophys 37:275–316
Soltani N, Khavari-Nejad RA, Yazdi MT, Shokravi S (2007) Growth and some metabolic features of cyanobacterium *Fischerella* sp. FS18 in different combined nitrogen sources. J Sci Repub Iran 18:123–128
Spolaore P, Joannis-Cassan C, Duran E, Isambert A (2006) Commercial applications of microalgae. J Biosci Bioeng 101:87–96
Sun L, Shumei W, Chen L, Gong X (2003) Promising fluorescent probes from phycobiliproteins. IEEE J Sel Top Quant 9:177–188
Takano H, Arai T, Hirano M, Matsunaga T (1995) Effects of intensity and quality of light on phycocyanin production by a marine cyanobacterium *Synechococcus* sp. NKBJ 042902. Appl Microbiol Biotechnol 43:1014–1018
Tomasseli L, Margheri MC, Sacchi A (1995) Effects of light on pigments and photosynthetic activity in a phycoerythrin- rich strain of *Spirulina subsalsa*. Aquat Microb Ecol 9:27–31
Tomasseli L, Boldrini G, Margheri MC (1997) Physiological behaviour of *Arthrospira* (*Spirulina*) maxima during acclimation to changes in irradiance. J Appl Phycol 9:37–43
Tsekos I, Niell FX, Aguilera J, Lopez-Fiueroa F, Delivopoulos SG (2002) Ultrastructure of vegetative gametophytic cells of *Porphyra leucosticta* (Rhodophyta) grown in red, blue and green light. Phycol Res 50:251–264
Tyagi R, Srinivas G, Vyas D, Kumar A, Kumar HD (1992) Differential effects of ultraviolet -b radiation on certain metabolic processes in a chromatically adapting *Nostoc* sp. Photochem Photobiol 55:401–407
Unsal-Kacmaz K, Makhov GJD, Sancar A (2002) Preferential binding of ATR protein to UV-damaged DNA. Proc Natl Acad Sci U S A 99:6673–6678
Verma K, Mohanty P (2000) Alteration in the structure of phycobilisomes of the cyanobacterium *Spirulina platensis* in response to enhanced Na? level. World J Microbiol Biotechnol 16:795–798
Xia J (2005) Response of growth photosynthesis and photoinhibition of the edible cyanobacterium *Nostoc sphaeroides* colonies to thiobencarb herbicide. Chemosphere 59:561–566
Zaccaro MC, Salazar C, Zulpa de Caire G, Storni de Cano M, Stella AM (2000) Lead toxicity in cyanobacterial porphyrin metabolism. Environ Toxicol 16:61–67
Zhou W, Tang X, Xiao H, Wang Y, Wang R (2009) Response of marine microalgae, heterotrophic bacteria and their relationship to enhanced UV-B radiation. Rev Ocean Coast Sea Res 8:35–38
Zucchi MR, Neechi O (2001) Effects of temperature, irradiance and photoperiod on growth and pigment content in some fresh water red algae in culture. Phycol Res 49:103–114

Advances in Production Technology 6

6.1 Introduction

Continuous growth of world population and fast-developing economy has led to a demand of biomass >50% in the upcoming two decades (Foley et al. 2011). This is a critical issue for finding abundantly biomass as carbon source in feedstock. The cultivation of cyanobacteria is one of the modern biotechnologies for production of biomass. Previously, mass culture of microalgae was the primary focus of research in USA, Germany, and Japan (Borowitzka 1999). Nevertheless, microalgal biomass is a critical commodity in most of agricultural crops yield, at a very low cost of production technology (Vandamme et al. 2013). Most of the microalgae species are profusely grown in aquatic environment in different elevation of CO_2 concentration, which indicates the potentiality of CO_2 mitigation strategies. They efficiently utilize light energy, macronutrients (carbon, hydrogen, nitrogen, oxygen, etc.), and micronutrients for making valuable biomass. A few microalgal species require organic carbon for growth through heterotrophy which shows more advantage over autotrophic culture (Perez-Garcia et al. 2011). The exclusive demand of microalgal biomass has increased the need of research and development for high production of biomass through photobioreactor (Morweiser et al. 2010), selection of productive strain (Larkum et al. 2012) and genetic engineering of metabolic pathway (Georgianna and Mayfield 2012). A few researches have been made on downstream processing essential to reduce cost of biomass production (Greenwell et al. 2010; Chen et al. 2010). There are few microalgae which are cultivated under certain stress condition that play an important role in growth and development of organism for production of some specific products for human welfare (Skjånes et al. 2012). Certain cyanobacteria such as *Spirulina* sp., *Nostoc* sp., etc. have been utilized for commercialized production in several countries for its potential application in health food (Dillon et al. 1995). Most of the microalgal species have inexpensive source of pigment known as phycobiliproteins (Table 6.1).

The various methodologies of cultivation types of PC/PE/APC production have been reported (Eriksen 2008; Kuddus et al. 2013). Thus, the production of

V.K. Kannaujiya et al., *Phycobiliproteins: Recent Developments and Future Applications*, https://doi.org/10.1007/978-981-10-6460-9_6

Table 6.1 Phycobiliproteins composition in different microalga

Microalga	Pigment	Amount (% in dry weight)	References
Anabaena sp.	Phycocyanin	8.3	Markou and Nerantzis (2013) and Sekar and Chandramohan (2008)
Nostoc sp.	Phycocyanin	20	Markou and Nerantzis (2013) and Sekar and Chandramohan (2008)
Phormidium valderianum	Phycocyanin	20	Markou and Nerantzis (2013) and Sekar and Chandramohan (2008)
Spirulina fusiformis	C-phycocyanin	46	Markou and Nerantzis (2013) and Sekar and Chandramohan (2008)
Rhodosorus marinus	Phycoerythrin	8	Markou and Nerantzis (2013) and Sekar and Chandramohan (2008)
Porphyridium cruentum	B-phycoerythrin/R--phycocyanin	32.7/11.9	Markou and Nerantzis (2013) and Sekar and Chandramohan (2008)
Spirulina platensis	C-phycocyanin/allophycocyanin	9.5/9.6	Markou and Nerantzis (2013) and Sekar and Chandramohan (2008)
Spirulina sp.	C-phycocyanin/allophycocyanin	17.5/20	Hifney et al. (2013)

microalgal biomass for collection of phycobiliproteins is of bright interest today. In this chapter, we have provided detailed mechanism of cultivation and strategies adopted for culture of microalgal species and efficient production of phycobiliproteins.

6.2 Strategies and Types of Cultivation

6.2.1 Mass Culture of Microalgae

In the past few years, there have been a number of researches concerned about the development of microalgae culture technology for large-scale production. The large-scale productions of microalgae are keenly dependent on the trophic level for their potential productivity. Among the various configuration concepts, there are few types of trophic level for growth of microalgae (Table 6.2).

6.2.1.1 Phototrophic

Photoautotrophic production is an outdoor method for microalgae production in open sunlight (Jiménez et al. 2003). Outdoor culture is widespread raceway system including artificial ponds and large containers for growth of microalgae. Outdoor

Table 6.2 Physiochemical parameters for different cultivations of microalga

Cultivation condition	Source of energy	Source of carbon	Cell density	Reactor	Cost	Issues
Phototrophic	Light	Inorganic	Low	Open pond or photobioreactor	Low	Low cell density
						High condensation cost
						Contamination
Heterotrophic	Organic	Organic	High	Conventional fermentor	Medium	Contamination
						High substrate cost
Mixotrophic	Light and organic	Inorganic and organic	Medium	Closed photobioreactor	High	High equipment cost
						High substrate cost
Photoheterotrophic	Light	Organic	Medium	Closed photobioreactor	High	High equipment cost
						High substrate cost

Adapted from Chen et al. (2011)

culture can be grown in natural lakes, lagoons, and ponds. The cyanobacterium *Arthrospira platensis* is one of the highly tolerable and luxuriant grown organisms in alkaline pH devoid of any contamination in open environment or outdoor culture. It is a highly abundant cyanobacterium in outdoor culture which reaches up to 3000 mt dry weight production worldwide annually and plays significant role in biomass production with value-added products. Recently, biomass productivities of photoautotrophic organism have been improved with involvement of various kinds of photobioreactors (Chen et al. 2011). Recently, certain literatures have been pointed out the types of photobioreactor and culture conditions for accumulation of phycocyanin in *Arthrospira platensis* (Zeng et al. 2012). Recently, certain types of closed bioreactors have designed for production of high-value food products also including pharmacy and cosmetic products (Xie et al. 2015).

6.2.1.2 Heterotrophic

The heterotrophic growth of microalgae is partially dependent on the light sources and shows higher growth rate as compared to light-dependent condition (Kuddus et al. 2013). The heterotrophic culture of microalgae is easier to scale up by reactor size/volume for obtaining better productivity. The unicellular rhodophyte *Galdieria sulphuraria* 074G is a potential organism for investigation and optimum productivity of phycocyanin in heterotrophic condition (Schmidt et al. 2005; Graverholt and Eriksen 2007; Kuddus et al. 2013). The cyanobacterium *Arthrospira* strain is also grown heterotrophically in glucose and fructose medium in darkness. However, the production of phycocyanin is quite low as compared to phototrophic condition (Muhling et al. 2005). The heterotrophic growth of *Galdieria sulphuraria* was considerably lower as compared to *Arthrospira platensis*; however, phycocyanin production was quite high (0.86 g L^{-1} day^{-1}) (Graverholt and Eriksen 2007).

6.2.1.3 Mixotrophic

The mixotrophic cultivation of microalgae is carried out in closed photobioreactors. Marquez et al. (1993) found optimum specific growth rate of mixotrophic cultures grown on glucose medium as compared to photoautotrophic culture. Thus, mixotrophic culture results in faster growth rate with higher biomass as compared to phototrophic culture (Vonshak et al. 2000; Chojnacka and Noworyta 2004). The productivities of phycocyanin contents are found more in the mixotrophic cultures in closed condition (Eriksen 2008; Kuddus et al. 2013).

6.2.2 Cultivation Technology

The commercial microalgae mass culture of *Chlorella* and *Spirulina* is now over 40–50 years old practice for being essential development of health food product. Nowadays, a wide variety of mass culture systems has been involved to culture in controlled/uncontrolled environmental conditions. Thus, an utmost need is to adapt microalgae under various physiological and biochemical parameters to scale up the culture from laboratory to large-scale production. Therefore, some improvements in

Table 6.3 Algae production companies around the world for various developments of bioproducts

Cultivation methods	Company	Country
Natural system	Kelco, San Diego	USA
Natural system	Neptune Industries, Boca Raton	USA
Natural system	Blue Marble Energy, Seattle	USA
Natural system	Aquaflow Bionomics, New Zealand	New Zealand
Open system	LiveFuels, Menlo Park, California	USA
Open system	OriginOil Inc., Los Angeles, California	USA
Open system	PetroSun, Scottsdale, Arizona	USA
Open system	Neste Oil, Helsinki	European
Open system	Ingrepo, Netherlands	European
Open system	Seambiotic, Ashkelon, Israel	Mediterranean
Closed system	A2BE Carbon Capture, Boulder, Colorado	USA
Closed system	GreenFuel Technologies, Cambridge, Massachusetts	USA
Closed system	Solazyme, Inc., San Francisco	USA
Closed system	Algenol Biofuels, Fort Meyers, Florida	USA
Closed system	Sapphire Energy, San Diego	USA
Closed system	Inventure Chemical Technology, Seattle	USA
Closed system	Solena, Washington State	USA
Closed system	Solix Biofuels, Fort Collins, Colorado	USA
Closed system	XL Renewables, Phoenix, Arizona	USA
Closed system	Bionavitas, Snoqualmie, Washington	USA
Closed system	Cellana, Hawaii	USA

For details see reference Singh and Gu (2010)

biophysical and economic feasibility are required for cultivation of microalgae in successive mass culture for production of high-value products. To obtain high-value products, microalgal biomass is gaining interest for cost-effective production. There are three types of cultivation systems characterized and named as natural, open, and closed by different worldwide companies used for production of microalgae (Table 6.3). The current development in microalgae production in EU countries has been widely reviewed. Similarly, an inclusive survey has conceded to highlight the various aspects of cultivation in different companies around the world (Fig. 6.1a). Most of the companies follow closed system of cultivation; however, few used suitable regions for open pond or natural systems (Fig. 6.1b).

6.2.2.1 Natural Systems

Natural system is the naturally created mass cultivation ponds of microalgae. In natural system, open ponds are usually uneven in size and depth. They seem like a raceway system for open culture pond, and circulation of liquid is occurring by means of atmospheric air. The natural system may mimic the way algae grow in their natural environment due to several stress conditions. Among the hundreds of species of microalgae, *Arthrospira platensis* is found in ancient natural lake and

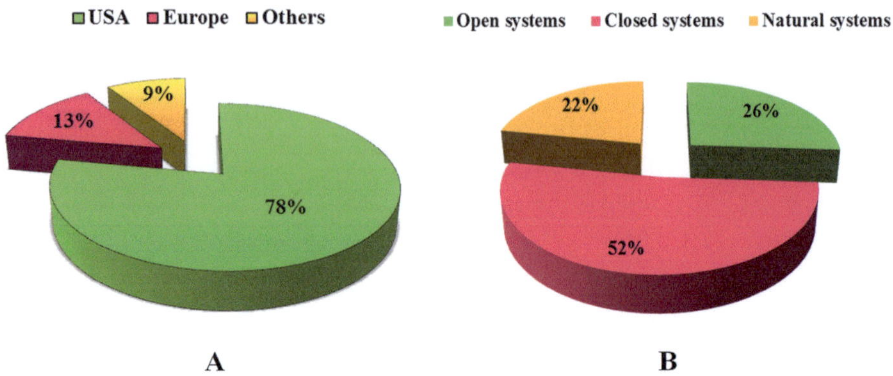

Fig. 6.1 Average percentage of companies around the world for producing of microalga and their products (**a**) and cultivation systems (**b**) (For details see reference Singh and Gu 2010)

Table 6.4 Natural lake of *Spirulina* sp. around the world

Natural lake	Regions	Country
Sambhar	Rajasthan	India
Texcoco	Mexico	USA
Chad	Chad	Africa
Bogoria	Kenya	Africa
Elementaita	Nairobi	Africa
Magadi	Kenya	Africa
Nakuru	Kenya	Africa
Twin Taung	Myanmar	Myanmar

adapts under invariably environmental condition. Here, we have described certain natural resources of *Spirulina* sp. around the world (Table 6.4).

6.2.2.2 Open Systems

The open systems are exclusively used for mass culture of microalgae. There are four major types of open-air systems such as shallow big ponds, tanks, circular ponds, and raceway ponds which are used efficiently for culture of microalga (Borowitzka 1999). The open systems directly interact with open environment; thus, cyanobacteria strongly face variable climatic condition. Furthermore, most of the companies have a very unique open system for large-scale production (Schlipalius 1991). Therefore, productivity of cyanobacteria is limited in open-air pond due to harsh environmental condition. To obtained high cell density, paddle-wheel-driven raceway ponds are used in Israel (Ben-Amotz 1995). Moreover, superfood cyanobacteria such as *Spirulina* and *Chlorella* grow well in a raceway system to achieve high production. In addition, depth and length of ponds are quite variable that reduces penetration of light along depth of water column. However, shallower ponds may induce the light penetration around algal cells; thus, there is an utmost need to optimize pond depth for optimum growth (Fig. 6.2).

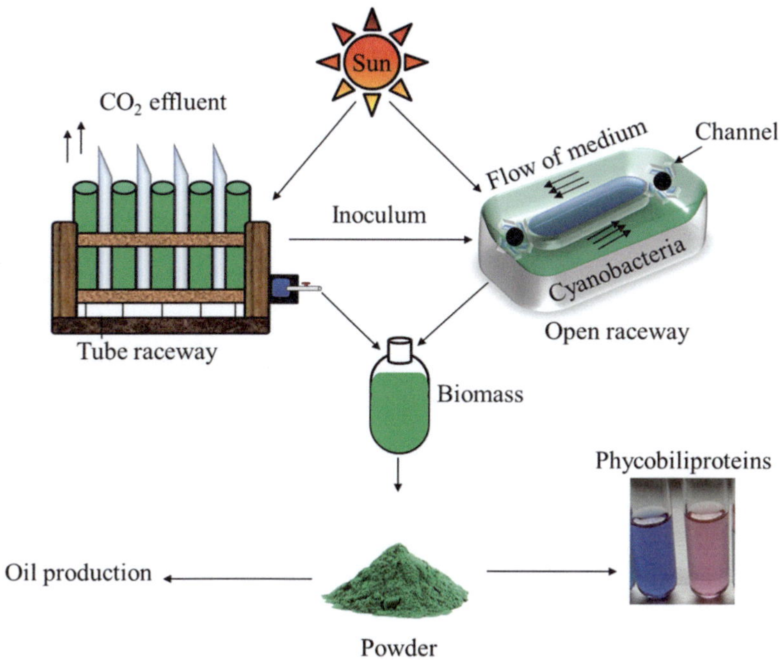

Fig. 6.2 Open photobioreactor raceway for large-scale production of microalgae and phycobiliproteins

The comparative relationship between depth of pond and inoculum density productivity has been extensively recorded for *Spirulina* sp. for large-scale production of biomass (Vonshak 1997). Borowitzka (Borowitzka 1999) reported preliminary calculations of open ponds (raceway) under warmer and sunnier condition for optimum growth of cyanobacteria. The calculation was optimized productivity up to 25 g^{-2} day^{-1} in semicontinuous mode throughout the year. In addition, cell concentrations were found 0.5 g/L in open ponds (Davis et al. 2011). The open pond systems require constant temperature (15 °C), less maintenance but produce lower yields of biomass. The photobioreactors are advancement of open raceway to produce more biomass in per unit area.

6.2.2.3 Closed Systems

Nevertheless, open system is more productive with minimum care, but variable environmental factor acts as limiting factors on growth of cyanobacteria. The closed photobioreactor has many advantages over open reactors in terms of monoseptic culture, prevention of water evaporation, and proper utilization of energy and chemicals that leads to higher productivity (Barbosa et al. 2003). Thus, optimum growth of microalga will require closed systems for mass culture with fixed environmental condition. Moreover, most of the cyanobacteria are not able to grow on highly selective environmental condition. Furthermore, potential applications of microalgae

Table 6.5 Diverse properties of different reactors for large-scale microalgae production

Reactor type	Mixing	Species control	Sterility	Scale-up	References
Shallow ponds	Poor	Difficult	None	Very difficult	Borowitzka and Borowitzka (1989)
Circular ponds	Good	Difficult	None	Very difficult	Soeder (1981)
Raceway ponds	Fair-good	Difficult	None	Very difficult	Oswald (1988) and Weissman and Goebel (1987)
Airlift	Good	Easy	Easy	Difficult	Jüttner (1977)
Flat plate	Excellent	Easy	Achievable	Difficult	Hu et al. (1996) and Tredici and Zitelli (1997)
Tubular	Excellent	Easy	Easy	Easy	Torzillo (1997)

For details see reference Borowitzka (1999)

products are keenly dependent on the purity of cyanobacteria; therefore, biotic and abiotic contamination may reduce the growth as well as the productivity of cyanobacteria. The idea of closed system has emerged for a long time ago (Jüttner 1977; Pirt et al. 1983). The growth of microalgae in closed system is ubiquitous either heterotrophically, mixotrophically, or photoautotrophically. Recently, several advancements have taken place in design and operation of closed photobioreactor to obtain high-end products by commercial production of microalgae. The closed photobioreactor has several advantages like clean culture, proper utilization of light, sustainable biomass, and temperature control factor that induce the productivity of microalgae (Chrismadha and Borowitzka 1994). The closed system requires minimum space to operate continuous culture mode for constant production of highly dense uniform biomass. These bioreactors require a constant source of CO_2 for production of biomass that plays an additional purpose of carbon sequestration. In recent years, there is a great advancement in design and operation of photobioreactors for large-scale commercial culture of cyanobacteria. Several designs of photobioreactors are being commercialized in near future. Moreover, all photobioreactors are designed to increase surface area and maximum utilization of light in CO_2 atmosphere for enhanced productivity. There are wide developments in microalgae cultivation in open and enclosed photobioreactors. The open bioreactors have low algal cell densities as compared to closed bioreactor. Devis et al. (Davis et al. 2011) found more (2–6 g/L) productivity in closed photobioreactors as compared to open ponds (0.5 g/L). Moreover, in outdoor open raceways, the PC productivity in cultures of *A. platensis* and *Anabaena* sp. have been found to be 14–23.5 and 0.82–1.32 g m^{-2} day^{-1}, respectively (Moreno et al. 2003). Therefore, PC, PE, and APC productivity may enhance up to tenfold in closed raceway bioreactor.

According to this fundamental principle, open photobioreactor is designed in several types and all dependent on size and environmental conditions. Closed photobioreactor is categorized into four major types such as plate, tube, annular, and airlift developed for commercialization and large-scale production of mass culture of cyanobacteria (Schenk et al. 2008) (Table 6.5; Fig. 6.3).

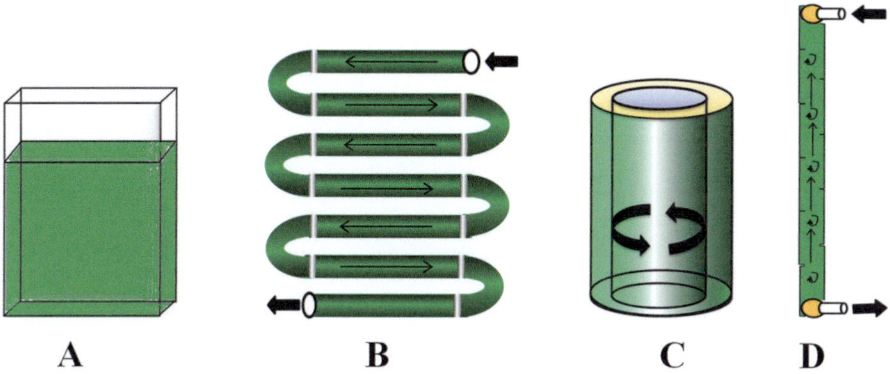

Fig. 6.3 Different units of closed photobioreactor. Plate (**a**), tubular (**b**), annular (**c**), plate airlift (**d**) (For details see the reference Schenk et al. 2008)

Some of closed photobioreactors are given in the following subsections which efficiently utilize the production of large-scale cyanobacteria and their bioproducts (oils and phycobiliproteins).

6.3 Photobioreactors and Its Utilization

6.3.1 Tubular Photobioreactor

Among the closed systems, tubular photobioreactors are widely used at the industrial level for production of microalgal biomass (Pulz et al. 2013). The advantages of tubular photobioreactors have been discussed in literature (Torzillo 1997; Tredici et al. 2010; Chini Zittelli et al. 2013; Torzillo and Zittelli 2015). The tubular bioreactor has been made up by an array of clear transparent tubes and widely used photobioreactor worldwide. The culture is continuously circulated inside the tubes for proper utilization of air and light for photosynthesis.

Tubular photobioreactors are subdivided to many types on variability of reservoir design (Table 6.6). The common design is represented in Fig. 6.4.

6.3.2 Flat Panel Photobioreactor

The flat panel photobioreactors have more advantages over tubular bioreactor due to its high surface-area-to-volume ratio. The flat panel bioreactors are connected by an array of flat reservoir plate which is made up of plate of polycarbonate with stainless steel. The circulation of air in flat panel bioreactor is achieved by either bubbling or perforated tube. Illumination of culture panel is achieved by incorporation of bright fluorescent tubes at one surface. This bioreactor is involved for more production of biomass as compared to tubular bioreactor. The unique geometry of fed-batch

Table 6.6 Types of tubular photobioreactors

Tubular photobioreactor	Design	References
Bubble photobioreactor	Array of elongated cylinder with high area–volume ratio	Sánchez Mirón et al. (2000)
Airlift photobioreactor	Two connected tubes: riser (gas phase) and downcomer (non-gas phase)	Sánchez Mirón et al. (2000), Tredici et al. (2010), and Molina Grima et al. (1999)
Serpentine photobioreactor	Glass U bends form loops, gas, and nutrient flow maintained by airlift mechanism	Torzillo et al. (1993), Acién Fernández et al. (2013), and Molina Grima et al. (2013)
Manifold photobioreactor	A series of parallel tubes connected by two manifolds for distribution and collection of biomass	Tredici et al. (2010) and Pulz et al. (2013)
Helical photobioreactor	Thin-diameter elastic tubes bound around and upright for supporting the structure of bioreactor	Tredici and Chini Zittelli (1998), Morita et al. (2000), and Raes et al. (2014)

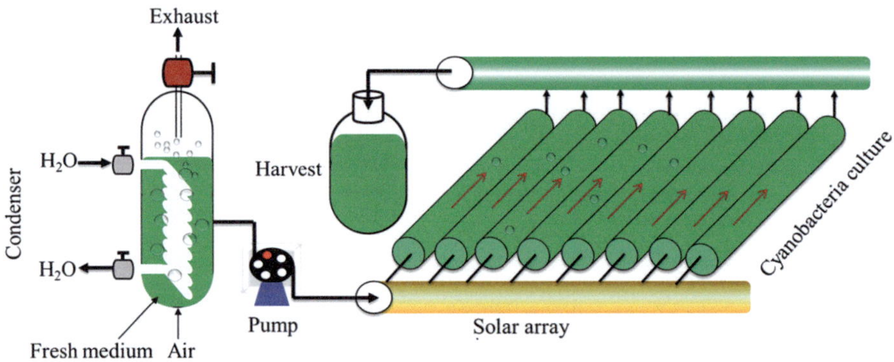

Fig. 6.4 Design of tubular photobioreactor

photobioreactor also influenced the volumetric productivity and final biomass concentration. Comparatively, flat panel bioreactors have 3.4-fold higher volumetric ratio than other photobioreactors and produced higher biomass (1.7 gDW L^{-1} day^{-1}) as compared to photobioreactors (0.5 gDW L^{-1} day^{-1}) (Bergmann and Trösch 2016).

6.3.3 Fed-Batch Photobioreactor

The fed-batch cultivation is an effective strategy to enhance in biomass of microalgae (Fig. 6.5). The relative production of PBPs may further enhance by 16.1% as compared to conventional production technology. However, more productivity was achieved (94.8 mg/L/d) using fed-batch strategy by incorporation of growth medium (5 mM) (Yang et al. 2008).

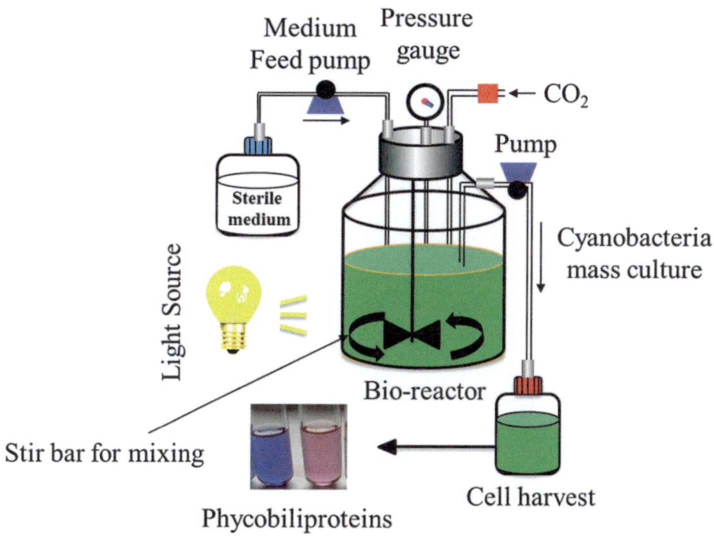

Fig. 6.5 Fed-batch photobioreactor for continuous production of biomass

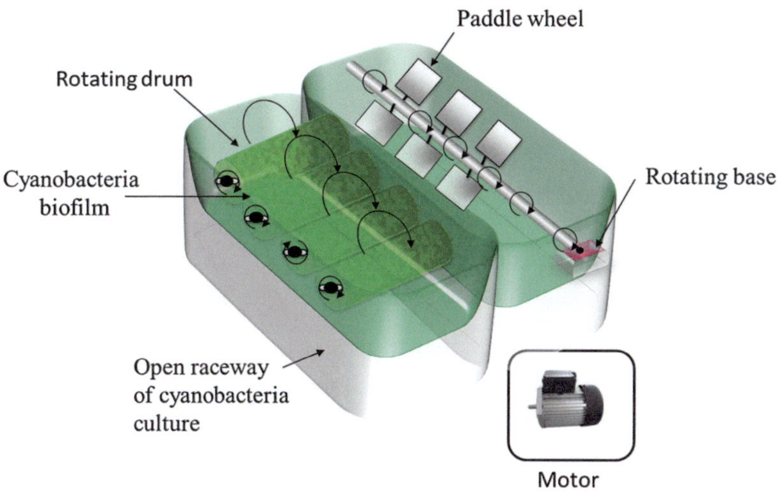

Fig. 6.6 Design of rotating algal biofilm photobioreactor

6.3.4 Rotating Algal Biofilm Photobioreactor

The commercial cultivation of microalgae production is still facing many challenges in operational and performance cost, processing of algal biomass (Fig. 6.6). The economic cost becomes more with low density microalgal growth suspension (Davis et al. 2011; Gross et al. 2013; Johnson and Wen 2010). To make a sustainable and cost-effective productivity, algal biofilm reactor has been developed. Historically, algal biofilm bioreactor was first reported in 1980 that was made up with rotating drum (polystyrene) for reduction of nitrogen and phosphorus inorganics from municipal waste water (Przytocka-Jusiak et al. 1984). Recently, same bioreactor has been designed for biofuel production (Christenson and Sims 2012). Nowadays, rotating algal biofilm bioreactor is utilized as a novel reactor for growth of microalgae and low-cost production of biomass for application in various fields of sciences (Christenson and Sims 2012; Gross et al. 2013). This photobioreactor having growth platform is coupled with water mobility (Gross et al. 2015) as sustainable growth medium which further reduces the costs of phycocyanin production (Xie et al. 2015). The productivity of phycocyanin was found lower in algal biofilm (820–850 mg/m^2-day) with higher purity (0.23 ± 0.03) as compared to open photobioreactor (1350 ± 173 mg/m^2-day) (Pushparaj et al. 1997; Jiménez et al. 2003). Thus, rotating algal biofilm bioreactor has proved with several advantages including sustainable biomass productivity, easy to harvest, enriched CO_2, and light utilization (Gross et al. 2013).

6.4 Conclusion

The various kinds of photobioreactors are involved for large-scale cultivation and production of microalgae. The open photobioreactor has several advantages such as low cost and high-volume mass culture. However, risk of contamination is a drawback for quality mass products in utilization in various fields. The possibility of limitation has been reduced by closed bioreactor for high quality of bioproducts. Recently, several tubular closed bioreactors have been designed for large-scale microalgae cultivation with automatic operation and highly controlled system for continuous production. However, large-scale production of value-added compounds in closed bioreactor is limited due to high investment of cost and energy (Borowitzka and Moheimani 2013; Tredici et al. 2015). Therefore, open bioreactors are good resources for large-scale production of high-value food products and cosmetics for biotechnology industries. Recently, algal biofilm bioreactor has been an alternate resource for reduced harvesting as compared to closed photobioreactor. However, the development of algal film bioreactor is still needed for production of high-value compounds. Alternatively, recombinant protein production is another technique for production of certain compounds which could be utilized in large-scale production of particular compounds. However, the production of these holo-protein PBPs is more challenging than the production of other recombinant proteins. Recently, the components of PBPs such as holo-PC α subunit (Guan et al. 2007), apo-PC β subunit (Weissman and Goebel 1987), and holo-APC α subunit (Zeng et al. 2012) were constructed in *E. coli* genome for production. Therefore, future biomass industries are utmost needed to overcome certain limitations such as cost of production, energy utilization, and imbalance of high-value food products for proper utilization of biomass for biotechnology industry.

References

Acién Fernández FG, Fernández Sevilla JM, Molina Grima E (2013) Photobioreactors for the production of microalgae. Rev Environ Sci Biotechnol 12:131–151

Barbosa MJ, Albrecht M, Wijffels RH (2003) Hydrodynamic stress and lethal events in sparged microalgae cultures. Biotechnol Bioeng 83:112–120

Ben-Amotz A (1995) New mode of *Dunaliella* biotechnology: two-phase growth for b-carotene production. J Appl Phycol 7:65–68

Bergmann P, Trösch W (2016) Repeated fed-batch cultivation of *Thermosynechococcus elongatus* BP-1 in flat-panel airlift photobioreactors with static mixers for improved light utilization: Influence of nitrate, carbon supply and photobioreactor design. Algal Res 17:79–86

Borowitzka MA (1999) Commercial production of microalgae: ponds, tanks, tubes and fermenters. J Biotechnol 70:313–321

Borowitzka LJ, Borowitzka MA (1989) Industrial production: methods and economics. In: Cresswell RC, Rees TAV, Shah N (eds) Algal and Cyanobacterial Biotechnology. Longman Scientific, London, pp 294–316

Borowitzka MA, Moheimani NR (2013) Algae for biofuels and energy. Developments in applied phycology. Springer, Dordrecht, p 285

Chen C-Y et al (2010) Cultivation, photobioreactor design and harvesting of microalgae for biodiesel production: a critical review. Bioresour Technol 102:71–81

Chen C-Y, Yeh K-L, Aisyah R, Lee D-J, Chang J-S (2011) Cultivation, photobioreactor design and harvesting of microalgae for biodiesel production: A critical review. Bioresour Technol 102:71–81

Chini Zittelli G, Biondi N, Rodolfi L, Tredici MR (2013) Photobioreactors for mass production of microalgae. In: Richmond A, Hu Q (eds) Handbook of microalgal culture: applied phycology and biotechnology, 2nd edn. Wiley, Oxford, pp 225–266

Chojnacka K, Noworyta A (2004) Evaluation of *Spirulina* sp. growth in photoautotrophic, heterotrophic and mixotrophic cultures. Enzym Microb Technol 34:461–465

Chrismadha T, Borowitzka MA (1994) Effect of cell density and irradiance on growth, proximate composition and eicosapentaenoic acid production of *Phaeodactylum tricornutum* grown in a tubular photobioreactor. J Appl Phycol 6:67–74

Christenson L, Sims R (2012) Rotating algal biofilm reactor and spool harvester for wastewater treatment with biofuels by-products. Biotechnol Bioeng 109:1674–1684

Davis R, Aden A, Pienkos P (2011) Techno-economic analysis of autotrophic microalgae for fuel production. Appl Energy 88:3524–3531

Dillon JC, Phuc AP, Dubacq JP (1995) Nutritional value of the alga *Spirulina*. World Rev Nutr Diet 77:32–46

Eriksen NT (2008) Production of phycocyanin-a pigment with applications in biology, biotechnology, foods and medicine. Appl Microbiol Biotechnol 80:1–14

Foley JA, Ramankutty N et al (2011) Solutions for a cultivated planet. Nature 478:337–342

Georgianna DR, Mayfield SP (2012) Exploiting diversity and synthetic biology for the production of algal biofuels. Nature 488:329–335

Graverholt OS, Eriksen NT (2007) Heterotrophic high-cell density fed-batch and continuous-flow cultures of *Galdieria sulphuraria* and production of phycocyanin. Appl Microbiol Biotechnol 77:69–75

Greenwell HC et al (2010) Placing microalgae on the biofuels priority list: a review of the technological challenges. J Roy Soc Interf 7:703–726

Grima ME, Fernández FGA, García Camacho F, Chisti Y (1999) Photobioreactors: light regime, mass transfer, and scale up. J Biotechnol 70:231–247

Gross M, Henry W, Michael C, Wen Z (2013) Development of a rotating algal biofilm growth system for attached microalgae growth with in situ biomass harvest. Bioresour Technol 150:195–201

Gross M, Mascarenhas V, Wen Z (2015) Evaluating algal growth performance and water use efficiency of pilot-scale revolving algal biofilm (RAB) culture systems. Biotechnol Bioeng 112:2040–2050

Guan X, Qin S, Su Z, Zhao F, Ge B, Li F, Tang X (2007) Combinational biosynthesis of a fluorescent cyanobacterial holo-α-phycocyanin in *Escherichia coli* by using one expression vector. Appl Biochem Biotechnol 142:52–59

Hifney AF, Issa AA, Fawzy MA (2013) Abiotic stress induced production of β-carotene, allophycocyanin and total lipids in *Spirulina* sp. J Biol Earth Sci 3:B54–B64

Hu Q, Guterman H, Richmond A (1996) A flat inclined modular photobioreactor for outdoor mass cultivation of photoautotrophs. Biotechnol Bioeng 51:51–60

Jiménez C, Cossıo BR, Labella D, Niell FX (2003) The feasibility of industrial production of *Spirulina* (*Arthrospira*) in southern Spain. Aquaculture 217:179–190

Johnson M, Wen Z (2010) Development of an attached microalgal growth system for biofuel production. Appl Microbiol Biotechnol 85:525–534

Jüttner F (1977) Thirty liter tower-type pilot plant for the mass cultivation of light- and motion-sensitive planktonic algae. Biotechnol Bioeng 19:1679–1687

Kuddus M, Singh P, Thomas G, Al-Hazimi A (2013) Recent developments in production and botechnological applications of c-phycocyanin. Biomed Res Int 2013:742859

Larkum AWD et al (2012) Selection, breeding and engineering of microalgae for bioenergy and biofuel production. Trends Biotechnol 30:198–205

Markou G, Nerantzis E (2013) Microalgae for high-value compounds and biofuels production: a review with focus on cultivation under stress conditions. Biotechnol Adv 31:1532–1542

Marquez FJ, Sasaki K, Kakizono T, Nishio N, Nagai S (1993) Growth characteristics of *Spirulina platensis* in mixotrophic and heterotrophic conditions. J Ferment Bioeng 76:408–410

Molina Grima E, Acién Fernández FG, Medina AR (2013) Downstream processing of cell mass and products. In: Richmond A, Hu Q (eds) Handbook of microalgal culture: applied phycology and biotechnology, 2nd edn. Wiley, Oxford, pp 26–309

Moreno J, Vargas MA, Rodriguez H, Rivas J, Guerrero MG (2003) Outdoor cultivation of a nitrogen-fixing marine cyanobacterium, *Anabaena* sp. ATCC 33047. Biomol Eng 20:191–197

Morita M, Watanabe Y, Saiki H (2000) Investigation of photobioreactor design for enhancing the photosynthetic productivity of microalgae. Biotechnol Bioeng 69:63–668

Morweiser M et al (2010) Developments and perspectives of photobioreactors for biofuel production. Appl Microbiol Biotechnol 87:1291–1301

Muhling M, Belay A, Whitton BA (2005) Screening *Arthrospira* (*Spirulina*) strains for heterotrophy. J Appl Phycol 17:129–135

Oswald WJ (1988) Large-scale algal culture systems (engineering aspects). In: Borowitzka MA, Borowitzka LJ (eds) Microalgal biotechnology. Cambridge University Press, Cambridge, pp 357–394

Perez-Garcia O, Escalante FME, de-Bashan LE, Bashan Y (2011) Heterotrophic cultures of microalgae: metabolism and potential products. Water Res 45:11–36

Pirt SJ, Lee YK, Walach MR, Pirt MW, Balyuzi HHM, Bazin MJ (1983) A tubular bioreactor for photosynthetic production of biomass from carbon dioxide: design and performance. J Chem Technol Biotechnol 33:35–58

Przytocka-Jusiak M, Baszczyk M, Kosinska E, Bisz-Konarzewska A (1984) Removal of nitrogen from industrial wastewaters with the use of algal rotating disks and denitrification packed bed reactor. Water Res 18:1077–1082

Pulz O, Broneske J, Waldeck P (2013) IGV GmbH experience report, industrial production of microalgae under controlled conditions: innovative prospects. In: Richmond A, Hu Q (eds) Handbook of microalgal culture: applied phycology and biotechnology, 2nd edn. Wiley, Oxford, pp 445–460

Pushparaj B, Pelosi E, Tredici MR, Pinzani E, Materassi R, Autotrofi M (1997) An integrated culture system for outdoor production of microalgae and cyanobacteria. J Appl Phycol 9:113–119

Raes EJ, Isdepsky A, Muylaert K, Borowitzka MA, Moheimani NR (2014) Comparison of growth of *Tetraselmis* in a tubular photobioreactor (Biocoil) and a raceway pond. Appl Phycol 26:247–255

Sánchez Mirón A, García Camacho F, Contreras Gómez A, Grima EM, Chisti Y (2000) Bubble-column and airlift photobioreactors for algal culture. AICHE J 46:1872–1887

Schenk PM, Thomas-Hall SR, Stephens E, Marx UC, Mussgnug JH, Posten C, Kruse O, Hankamer B (2008) Second generation biofuels: high-efficiency microalgae for biodiesel production. Bioenergy Res 1:20–43

Schlipalius L (1991) The extensive commercial cultivation of *Dunaliella salina*. Bioresour Technol 38:241–243

Schmidt RA, Wiebe MG, Eriksen NT (2005) Heterotrophic high cell-density fed-batch cultures of the phycocyanin producing red alga *Galdieria sulphuraria*. Biotechnol Bioeng 90:77–84

Sekar S, Chandramohan M (2008) Phycobiliproteins as a commodity: trends in applied research, patents and commercialization. J Appl Phycol 20:113–136

Singh J, Gu S (2010) Commercialization potential of microalgae for biofuels production. Renew Sust Energ Rev 14:2596–2610

Skjånes K, Rebours C, Lindblad P (2012) Potential for green microalgae to produce hydrogen, pharmaceuticals and other high value products in a combined process. Crit Rev Biotechnol 1–44

Soeder CJ (1981) Productivity of microalgal systems. UOFS Publ Ser C 3:9–15

Torzillo G (1997) Tubular reactors. In: Vonshak A (ed) *Spirulina platensis (Arthrospira),* physiology, cell biology and biotechnology. Taylor and Francis, London, pp 101–115

Torzillo G, Zittelli GC (2015) Tubular photobioreactors. In: Prokop et al (eds) Algal biorefineries. Springer, Switzerland, pp 187–212

Torzillo G, Carlozzi P, Pushparaj B, Montaini E, Materassi R (1993) A two-plane tubular photobioreactor for outdoor culture of *Spirulina*. Biotechnol Bioeng 42:891–898

Tredici MR, Chini Zittelli G (1998) Efficiency of sunlight utilization: tubular versus flat photobioreactors. Biotechnol Bioeng 57:187–197

Tredici MR, Zitelli GC (1997) Cultivation of *Spirulina* (*Arthrospira*) *platensis* in flat plate reactors. In: Vonshak A (ed) Spirulina platensis (Arthrospira): physiology, cell biology and biotechnology. Taylor and Francis, London, pp 117–130

Tredici MR, Chini Zittelli G, Rodolfi L (2010) Photobioreactors. In: Flickinger MC, Anderson S (eds) Encyclopedia of industrial biotechnology: bioprocess, bioseparation, and cell technology. Wiley, Hoboken, pp 3821–3838

Tredici MR, Bassi N, Prussi M, Biondi N, Rodolfi L, Chini Zittelli G, Sampietro G (2015) Energy balance of algae biomass production in a 1-ha "Green Wall Panel" plant: how to produce algae biomass in a closed reactor achieving a high Net Energy Ratio. Appl Energy 154:1103–1111

Vandamme D, Foubert I, Muylaert K (2013) Flocculation as a low-cost method for harvesting microalgae for bulk biomass production. Trends Biotechnol 31:233–239

Vonshak A (1997) Outdoor mass production of Spirulina: the basic concept. In: Vonshak A (ed) Spirulina platensis (Arthrospira): physiology, cellbiology and biotechnology. Taylor and Francis, London, pp 79–99

Vonshak A, Cheung SM, Chen F (2000) Mixotrophic growth modifies the response of *Spirulina* (*Arthrospira*) *platensis* (Cyanobacteria) cells to light. J Phycol 36:675–679

Weissman JC, Goebel RP (1987) Design and analysis of microalgal open pond systems for the purpose of producing fuels. Solar Energy Research Institute, Report SERI: STR-231-2840, p 1–214

Xie Y, Jin Y, Zeng X, Chen J, Lu Y, Jing K (2015) Fed-batch strategy for enhancing cell growth and C-phycocyanin production of *Arthrospira (Spirulina) platensis* under phototrophic cultivation. Bioresour Technol 180:281–287

Yang X, Kuk J, Moffat K (2008) Crystal structure of *Pseudomonas aeruginosa* bacteriophytochrome: photoconversion and signal transduction. Proc Natl Acad Sci U S A 105:14715–14720

Zeng X, Danquah MK, Zhang S, Zhang X, Wu M, Chen XD, Ng IS, Jing K, Lu Y (2012) Autotrophic cultivation of *Spirulina platensis* for CO_2 fixation and phycocyanin production. Chem Eng J 183:192–197

Advances and Strategies of Purification Technology

7

7.1 Introduction

Cyanobacteria mainly consist of three light-harvesting pigments, namely, Chl-*a*, carotenoids, and PBPs (Reis et al. 1998; Singh et al. 2009). PBP consists of linear tetrapyrrole prosthetic groups called bilins that are found in cyanobacteria, red algae, and cryptomonads (Bermejo et al. 2002). The bilin prosthetic group is a close resemblance to bile pigment biliverdin, which serves as primary biosynthetic precursor (Glazer 1985). These prosthetic groups are classified into four groups such as phycocyanobilin, phycoerythrobilin, phycourobilin, and phycobiliviolin. According to light-harvesting properties, the PBPs in cyanobacteria are commonly categorized into four groups: (1) phycoerythrin (PE; λ_{max} = 490–570 nm), (2) phycocyanin (PC; λ_{max} = 610–625 nm), (3) phycoerythrocyanin (PEC; λ_{max} = 560–600 nm), and (4) allophycocyanin (AP; λ_{max} = 650–660 nm) (Sidler 1994). Certain cyanobacteria consist of 50 % PBPs of total soluble protein (Muramatsu and Hihara 2012). PBPs are water-soluble, brilliantly colored proteins and form multichain holo-protein structure (Singh and Montgomery 2013). The pathway of energy transfer within subunits of PBPs starts from PE to PC and reaches to core unit APC and finally donates photonic energy to Chl *a* for photosynthesis (Bermejo et al. 2002; Parmar et al. 2011; Johnson et al. 2014; Kannaujiya et al. 2016). About 20 % constitute of PC and APC are located at periphery and core region of the photosynthetic membrane (Ajayan et al. 2012; Singh and Montgomery 2013). In addition, PE is located at distal region of PBPs and plays an important role in the adaptation of harsh environmental changes (Rodríguez et al. 1991). The prominent composition of PC and APC (3:1) is found in *Spirulina platensis* that play a distinguished role in the harvesting of photonic energy from the sun (Cohen 1997). Each PBP is made up of heteromonomer of two subunits α and β. The α subunits (12–20 kDa) are comparatively smaller than β subunits (15–22 kDa) (Sinha et al. 1995; Kannaujiya and Sinha 2016a). Each unit of PBPs has specific absorbance and fluorescence emission maxima in the visible range of light. Therefore, an arrangement of chromophores within the PBPs may allow photon absorption and unidirectional transfer of light energy to

V.K. Kannaujiya et al., *Phycobiliproteins: Recent Developments and Future Applications*, https://doi.org/10.1007/978-981-10-6460-9_7

PSII in chlorophyll. Efficient energy transfers from PBPs are restricted up to 95 % due to its geometrical arrangement. Electrostatic interaction between basic polypeptides and acidic polypeptides is stabilizing assembly of PBPs and also facilitates the unidirectional energy flow toward photosynthetic reaction center (Tandeau de Marsac and Cohen-Bazire 1977; Bryant et al. 1979; Glauser et al. 1992). Spectroscopic and structural properties of PBPs exhibit several unique qualitative and quantitative features (Glazer 1984; Silman et al. 1999; Sun et al. 2009; Richa et al. 2011). PBPs have unique structural and functional properties; thus it could play an indispensable significance in biotechnological industry. They have wide therapeutic significance and could be used as antimicrobial, antioxidant, neuroprotective, hepatoprotective, and anti-inflammatory including food, cosmetics, and fluorescent dye for human welfare. However, the wide applications of PBPs are strongly dependent on purity index to the development of quality products. The process of extraction and purification of PBPs are immensely complex and adequate production of PBPs also limited to certain cyanobacterial strain. In the past few years, several physical and chemical methods for cell disruption have been adopted for improved cell extraction from any cyanobacteria (Sekar and Chandramohan 2008). Similarly, several techniques have been adopted for purification of PBPs from cyanobacteria and red algae; however, major techniques are limited to laboratory scale. Recently, large-scale purification technologies have been developed for high production of PBPs from cyanobacteria (Juin et al. 2015; Esquivel-Hernández et al. 2016; Sonani et al. 2016; Luo et al. 2016; Grilo et al. 2016). In this book chapter, we critically review the current development of extraction and purification technology for purification of PBPs in large scale. The analyses are also identifying market value and aspects of commercial utility for biotechnological industry.

7.2 Techniques of Purification

7.2.1 Extraction of Phycobiliproteins

Extraction of PBPs from cyanobacterial cells is very difficult due to the small size of vegetative cells and extremely thick cell wall with multilayered structure (Stewart and Farmer 1984; Wyman 1992). Several physical and chemical methods have been developed to disrupt thick cell wall and effective extraction of PBPs. Although, no standard technique has been found for quantitative extraction of microalgae pigments (Wiltshire et al. 2000). The strategy of extraction was not homogeneous from one cyanobacterium to another (Ranjitha and Kaushik 2005). Literally, to obtain higher yield of PBPs, the process of extraction can be divided into two stages. In the first stage, water-soluble PBP pigments have been isolated from cell-free extract treated as crude extract and purification of crude extract to obtain pure PBPs follow the second stage. Many conventional physical and chemical methods have been developed for efficient extraction of PBPs.

Table 7.1 Physical methods for extraction of PBPs from cyanobacteria

Methods	References
Ultrasonication	Bermejo et al. (2006) and Horváth et al. (2013)
French press cell	Bryant et al. (1976) and Sinha et al. (1995)
Liquid nitrogen	Stewart and Farmer (1984)
Freeze–thaw (−20 to 4 °C)	Sinha et al. (1995) and Kannaujiya and Sinha (2016a, b)
Nitrogen cavitation	Viskari and Colyer (2003) and Simpson (2010)
Microwave-assisted extraction (MAE)	Chemat et al. (2012), Juin et al. (2015), and Esquivel-Hernández et al. (2016)

7.2.1.1 Physical Method

Several physical methods have been applied for extraction of PBPs. Some of the important methods are described below (Table 7.1).

Sonication

Ultrasonic treatment is a process of disruption of cell wall by generation of cavitation in super sound wave or ultrasound water bath. This method is more advantageous for easy operation, short time, minimum loss, and high yield for PBPs. The intensity and time of sonication is most important factor for breakage of different types of cell wall in cyanobacteria (Benedetti et al. 2004). Earlier, sonication has been commonly used for extraction of *Porphyridium cruentum* and *Synechococcus* sp. (Vernet et al. 1990; Roman et al. 2002). In present concern, sonication is widely used for extraction of PBPs. Probably, sonication intensity has been fixed in the ranges of 20–50 KHz for better yield, but time of sonication depends on cell wall complexity. Sonication with fine sand particles is more helpful for extraction of PBPs by reducing time periods (Wiltshire et al. 2000). In addition, the addition of glass pearls in biomass before sonication (50 kHz) in the ratio of 1:1.1 (biomass/glass pearls) is extra advantageous over sand particle (Moraes et al. 2011a).

French Press Cell

French pressure cell is a pressure-built apparatus that applies very high pressure (1000–10,000 psi) to disrupt thick cell wall and plasma membrane for releasing of desired proteins. In laboratory scale, initially culture was partially homogenized in phosphate buffer and subjected to French pressure cell (Aminco Corp.) operating at 12,000 psi pressure (Thomas and Passaquet 1999). Similarly, Sinha et al. (1995) also reported that isolation of PBPs was very easy after disruption of cyanobacterial cells by French press cell.

Freeze–Thaw Cycles

The extraction of PBPs from cyanobacterial biomass by various techniques has been discussed in many literatures. Among several techniques of extraction, freezing and thawing technique is more advantageous and considered to be better method for obtaining higher purity. This method is more reliable in terms of quick response, high reproducible, and robustness. Consequence of freeze–thaw cycles of cyanobacterial biomass under −20 to 4 °C (Viskari and Colyer 2003) is effective technique for easy cell disruption (Stewart and Farmer 1984). The freezing of 1 h duration of time is enough to break the cell wall; however, no significant difference in yield was recorded after repeated freeze–thaw (Moraes et al. 2011a). Prior to freeze–thaw, cell disruption with the help of mortar and pestle is a more effective strategy for efficient extraction of PBPs (Kannaujiya and Sinha 2016a). In addition, sonication prior to freeze–thaw process is another appropriate method for efficient extraction of PBPs (Kannaujiya and Sinha 2016b). Wood et al. (2015) used lyophilized powdered biomass for extraction of PC through repeated freeze–thaw cycles.

Nitrogen Cavitation

Nitrogen cavitation is an easy method for cell disruption with requirement of fewer techniques as compared to other extraction protocol of PBPs (Viskari and Colyer 2003). Viskari and Colyer (2003) has been shown minimum time required for fast cell disruption and release of PBPs. Recently, Simpson (2010) has developed advance nitrogen cavitation technique by incorporation of pressurized vessel for nitrogen-induced disruption of cells. In details, cells were filled in oxygen-free pressure vessels where large quantity of nitrogen atom dissolved under very high pressure (5500 kPa). When pressure volumes are released suddenly, nitrogen make bubbles and rupture the cell membrane and release cellular components.

Mechanical Shear

The mechanical shear is an old method for breaking cell wall for extraction of PBPs. In the process of mechanical shear, raw material of cyanobacteria is subjected to high-speed machine for breakage of cell wall. The efficiency of extraction from cyanobacteria and purity of PBPs are not greatly improved as compared to previous methods. The mechanical shear methods are more suitable for large-scale primary extraction of PBPs.

Microwave-Assisted Extraction

Microwave-assisted extraction (MAE) is a unique automated sustainable technology for efficient extraction of PBPs by reducing time and energy (Chemat et al. 2012). This technique has been applied for fast extraction (eight- to tenfold) of PBPs from dried vegetative cells of *Porphyridium purpureum* (Juin et al. 2015).

7.2.1.2 Chemical Methods

Total extraction of PBPs from any cyanobacteria is not possible by only involvement of physical approach. The combination of physical and chemical methods is a probable approach used for proper extraction of PBPs from any cyanobacterial filaments. In present concern, several chemicals have been introduced for disruption of cell wall and inhibition of protease degrading enzyme that degrades PBPs in vitro condition.

Table 7.2 Chemical methods for the extraction of phycobiliproteins from cyanobacteria

Chemicals	References
25 mM PB, 10 mg/ml lysozyme (pH 7)	Kilpatrick (1985)
1% cellulase, 0.1% pectolyase	Viskari and Colyer (2003)
0.75 M PB (pH 7), 0.1–1.0% Triton X-100	Sinha et al. (1995)
50 mM PB, 1 mM PMSF, 10% (w/v) EDTA and 5% (w/v) sucrose	Kannaujiya and Sinha (2016a, b)
Acetate buffer, 2 mg/ml lysozyme (Triton X-100)	Viskari and Colyer (2003)
250 mM Trizma, 10 mM EDTA	Viskari and Colyer (2003)
1% Rivanol	Minkova et al. (2003)
1 M Tris (pH 8.0), 0.5 M EDTA, 20% sucrose (w/v), 5 mg/ml lysozyme	Bhaskar et al. (2005)
1 M PB and NP-40	Zhao et al. (2015)

PB phosphate buffer, *PMSF* phenylmethanesulfonyl fluoride, *EDTA* ethylenediaminetetraacetic acid

Lysozyme

Lysozymes contain hydrolase enzymes (N-acetylmuramide glycanhydrolase) which catalyze hydrolysis of 1-4-beta linkages between N-acetylmuramic acid and N-acetyl-D-glucosamine in peptidoglycan structure. Therefore, structure of cyanobacterial cell wall was damaged and released PBPs. In addition, lysozyme-induced disintegration/degradation of cell wall followed by cellular fractionation could be more helpful as compared to crude extraction (Boussiba and Richmond 1979). Lysozyme was used with combination of 100 mM sodium phosphate buffer (pH 7.0) and 100 mM sodium EDTA for extraction of PBPs from cyanobacteria at room temperature (Boussiba and Richmond 1980).

Hydrochloric Acid

Hydrochloric acid (2–10 N) was used for extraction of PBPs from sheathed cyanobacteria in 50 mM phosphate buffer (pH 6.8) (Sarada et al. 1999). The composition of PBP recovery is insignificant from non-sheathed cyanobacteria. Hydrochloric acids are highly toxic and damaging in nature; therefore major components of PBPs get damaged and could not be recovered after extraction.

Detergents

The combination of 1% Triton X and 5% (w/v) sucrose is very competent mixture for extraction of PBPs from cyanobacteria without change in quality of proteins (Sinha et al. 1995). NP-40 is another potent surfactant that is used for rapid extraction of PBPs in any cyanobacteria (Zhao et al. 2015). Several combinations have been developed for easy extraction of PBPs (Table 7.2).

Extraction Mixture

The chemical compositions of 5–10 % EDTA (w/v), 5–10 % sucrose (w/v), 2.5 mM PMSF, and sodium azide are widely used for extraction PBPs with any physical methods (Sinha et al. 1995; Wang et al. 2014; Kannaujiya and Sinha 2015;

Kannaujiya and Sinha 2016a, b). This method is most appropriate for extraction from any cyanobacteria without loss of PBPs. Certain physical properties such as temperature and pH may affect extraction process during extraction of PBPs. The pH of solvents in distilled water (pH 7.0), seawater (pH 8.13), and 0.1 M phosphate buffer (pH 6.8) has a probable role in extraction of PBPs at room temperature (Sudhakar et al. 2015). Temperature (30–60 °C) is another factor for enhancement in extraction of PBPs from any cyanobacteria under optimum pH (6.8–7.5) condition.

7.2.2 Purification of Phycobiliproteins

During the past decades, separation and purification technology have been contributed significantly for advances in pharmaceutics and biotechnological industries. The various conventional separation methods including filtration, precipitation, electrophoresis, gel permeation, and column chromatography have been utilized as reliable and effective tools for rapid purification and enrichment of quality of PBPs. A few methods have been developed for large-scale purification and downstream processing of PBPs for conventional application. For small scale, there is a wide variety of column chromatography which has been developed for purification of PBPs (Sonani et al. 2016). Some of the conventional methods and column chromatography are described here for better understanding purification technique.

7.2.2.1 Precipitation and Dialysis

The solubility of proteins dramatically rises upon increasing in ionic strength of salt (<0.15 M); this phenomena is termed as salting-in. Beyond a permissible limit, increasing ionic strength of salt decreases solubility of proteins leading to precipitation in solution; this effect is termed as salting-out (Green and Hughes 1955). The mechanism of salting-out is dependent on preferential exclusion of solvent from surface water molecules closely associated with proteins. Hydration layer of water molecules (0.3–0.4 g water per gram protein) plays a critical role to maintaining solubility of globular proteins in native conformation (Rupley et al. 1983). Structurally, water molecules are interacted with polar, ionic, and hydrophobic amino acids of globular proteins. Hydrophobic interaction between amino acids is relatively weak due to low entropy and energy unfavorable. The surface tension of water molecules is increased when concentration of solvent increases facilitates more hydrophobic interaction. Thus, proteins minimize its size for reducing surface area and precipitation against salt effects (Wingfield 2001). The precipitation of proteins by using ammonium sulfate is widely adopted methods for partial purification of PBPs. The blue supernatant of PC was fractionally precipitated with 20–60 % (w/v) ammonium sulfate. Crude extract was treated with 20 % (w/v) ammonium sulfate for initial precipitation of large molecules. Thereafter, subsequent solution is treated with 60 % (*w/v*) ammonium sulfate in 50 mM Na-phosphate buffer (pH 7) for precipitation of PC (Kannaujiya and Sinha 2016a, b). The precipitated solution was subjected to dialysis via dialysis tubing (<12 kDa) against the same buffer to remove salt. The polar solvent like acetone has also been used for precipitation of

PE from crude extract of cyanobacterium *Lyngbya arboricola* (Tripathi et al. 2007). In this method, crude extract of PC was precipitated by 55 % of acetone for removal of small biomolecules, and subsequently precipitated sample was again subjected to 40 % for removal of carbohydrate, lipid, and proteins.

7.2.2.2 Expanded Bed Adsorption Chromatography

To obtain purified protein biomolecules, crude sample channelized by downstream processing involves various kinds of process including centrifugation, precipitation, dialysis, partition chromatography, and crystallization. This is economically, physically huge burden on sustainable production and low market value. To reduce the cost of bioproducts, expanded bed adsorption (EBA) has been developed in the early 1990s, and the term expanded bed was coined by Chase and Draeger (1992). This technique has multifunctional operation in a single unit which simplifies existing downstream process by uniting together three classical processes like contamination removal, concentration enhancement, and primary purification (Pierce et al. 1999; Anspach et al. 1999; Gonzalez et al. 2003). EBA chromatography is based on adsorption of desired proteins on solid surface and clarification of contaminant by mobile phase (liquid). Many types of beads on solid matrix have been developed including agarose microporous matrix on quartz surface with anion ligands (Streamline DEAE, GE Healthcare), agarose/cellulose microporous matrix on glass or tungsten carbide (Streamline Direct, GE Healthcare), and nonporous polymer with functional ligand-coated matrix on silica or zirconia ceramic core (HyperD; BioSepra, Life Technologies) (Tsai et al. 2004; Li et al. 2014). Streamline DEAE has been widely studied for adsorption isotherms and found an indispensable tool for separation and purification of proteins (Bermejo et al. 2006; Ramos et al. 2010, 2011; Bermejo and Ramos 2012). The details about kinetic behavior and adsorption isotherm in response to temperature and pH for Streamline Q XL (Moraes et al. 2011b) and Streamline DEAE further verify and establish optimum adsorption equilibrium for PC purification (Sala et al. 2014).

7.2.2.3 Gel Filtration Chromatography

Gel filtration chromatography is a kind of size exclusion chromatography to separate biomolecules in a native state by segregation of different molecular weights in non-denaturing condition (Fig. 7.1a).

This technique is relatively simple and also known with different name such as gel permeation, gel exclusion, molecular sieve, and partition chromatography. The separation of molecules is based on partition coefficient between stationary support (solid) and mobile phase (liquid). The gel filtration media is composed of spherical beads which form by modification of polysaccharides and resin. The spherical beads have network of pores around themselves for separation of biomolecules. There are some large molecules that are not able to enter into the pore called exclusion limit of beads, while small molecules easily enter to pores and retardate flow rate of molecules. Therefore, separation of mixture of molecules on gel filtration chromatography occurs in order of decreasing in size. Monomeric PE (44 kDa) and PC (34 kDa) have been obtained after gel filtration chromatography (Kannaujiya and Sinha 2016a) (Fig. 7.1b).

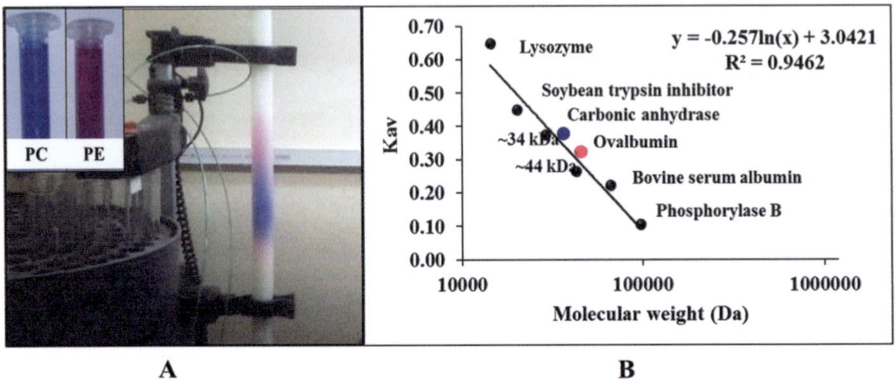

Fig. 7.1 The process of gel filtration chromatography in fast protein liquid chromatography (Akta Prime Plus, GE Healthcare, Uppsala, Sweden) (**a**) and calculation of molecular weight of PC and PE with different proteins marker in semilogarithmic plot (**b**) (Adapted and modified from Kannaujiya and Sinha 2016a). Inset shows purified PC (blue) and PE (pink)

In present concern, gel filtration chromatography is made up of dextran, allyl dextran bis-acrylamide, and agarose. In industry, Sephadex G-10 and G-25 column is made from dextran and used for desalting of proteins. Another, gel filtration column such as Sepharose and Sephacryl is made up of agarose and allyl dextran bis-acrylamide, respectively, which is used for separation of biomolecules. Kannaujiya and Sinha (2016a) found that monomeric molecular weight of PC and PE was separated and purified in gel filtration column of Sephacryl S-100 HR (1–100 kDa; 16/60, GE Healthcare, Uppsala, Sweden). Several cycles of gel filtration chromatography greatly improved purity index from 1.1 to 6.3 and 0.92 to 1.3 for PE and PC, respectively (Kannaujiya and Sinha 2016a). The surprising enhancement of PE purity index was may be due to monomeric nature and several fractionation cycles. Similarly, fractionation of PE with Sephacryl S-300HR gel filtration was also enhancement of purity index from 2.57 to 3.16. Therefore, this column is also used to purify complex mixture of biomolecules (Fig. 7.2).

7.2.2.4 Hydrophobic Interaction Chromatography

The protein separation is critically dependent on their molecular weight, biospecific characteristics, hydrophobicity, and net charge (Garcia and Pires 1993). The hydrophobic interaction chromatography (HIC) promotes hydrophobicity between immobilized stationary phase and nonpolar side chain of proteins.

The name HIC was introduced by Hjertén (1973) and demonstrated the separations of proteins on hydrophobic carbohydrate gel matrices by using ionic mobile phase. The adsorption of sample increases with increasing salt concentration in mobile phase, and pure adsorb sample is eluted by reduction in salt concentration (Fausnaugh and Regnier 1986; Roe 1989). The probable reason of adsorption of sample may induce by mechanism of van der Waals forces. HIC is an alternative technique that considered hydrophobic nature of protein residue with weak-binding

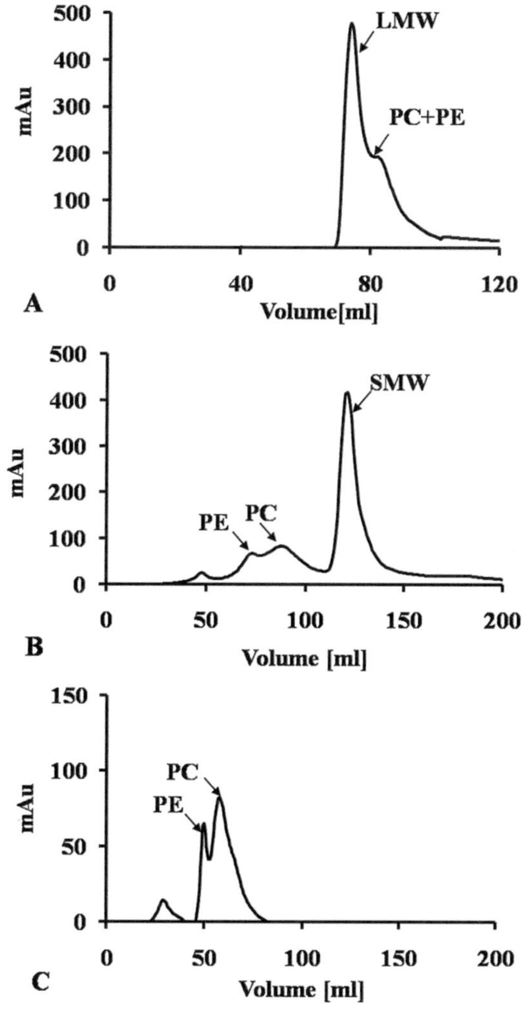

Fig. 7.2 Separation of PC and PE component of PBPs by gel filtration chromatography (Sephacryl S-100 HR) in the absence of NaCl (**a**), presence of NaCl (**b**), and eluted partially purified fractions of PC and PE (**c**) (Adapted from Kannaujiya and Sinha 2016a). *LMW* large-molecular-weight proteins, *SMW* small-molecular-weight proteins (Adapted from Kannaujiya and Sinha 2016a)

interaction by van der Waals which significantly reduces damage of proteins as compared to other conventional chromatography. In present concern, HIC is a major and alternative technique for separation and purification of proteins at laboratory as well as large scale. In the past few decades, a large variety of stationary phases for HIC has been promoted for separation and purification of diverse kind of biomolecules such as nucleic acids, hormones, enzyme, and recombinant proteins (Angelova et al. 1997; Queiroz et al. 1999, 2001). Moreover, the immobilized microporous surface of the stationary phase is functionalized by the interaction of nonionic hydrocarbon tails including octyl, butyl, hexyl, or phenyl ligands which are probable sites for reversible binding of proteins. Kannaujiya and Sinha (2016a) purified PC and PE by using phenyl FF HIC column chromatography (GE Healthcare, Uppsala, Sweden) in the presence of non-chaotropic salt $(NH_4)_2SO_4$ for the efficient

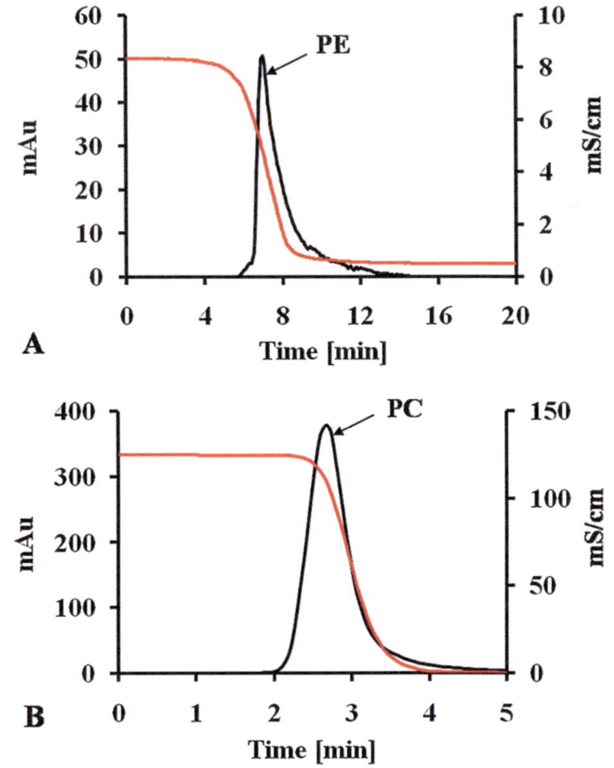

Fig. 7.3 Separation and purification of PE (**a**) and PC (**b**) from the isolated fractions of gel filtration chromatography by HiTrap Phenyl FF HIC at a flow rate of 60 mL h^{-1} with a linear ionic strength gradient (red line) of ammonium sulfate (1.5 M) in 50 mM PB (pH 8.5) (Adapted from Kannaujiya and Sinha 2016a)

hydrophobic binding and release of purest form of PC and PE after desalting (Fig. 7.3). Soni et al. (2008) used 20 mM Tris-Cl buffer and 1.5 M $(NH_4)_2SO_4$ (pH 8.1) for pre-equilibration of methyl macro-prep HIC column (Bio-Rad, USA) for lading of sample. A step gradient has been prepared with 0.25–1.5 M $(NH_4)_2SO_4$ for elution of bound proteins and higher purity of PC (3.85) and APC (7.49) obtained.

7.2.2.5 Ion Exchange Chromatography

Ion exchange chromatography (IEC) is used for separation and purification of ionic/charge molecules due to differences in charge properties. This chromatography is widely applicable for separation of various biomolecules. The versatile application is due to moderate cost, high resolving ability, easy scale-up, easy handling, and broad application in different sectors of biotechnology. IEC also consists of two-phase solid and liquid for ionic phase separation of biomolecules. They are found in two forms such as anion exchanger and cation exchanger for purification of sample. The exchangeable matrix consists of hydroxyl group, hydrogen group, and mono/diatomic, polyatomic, organic, basic, and acid ions. During separation of sample through IEC, mobile phase pH plays a critical role for ion exchanger through matrix. The IEC charge can be divided into two groups: positively charged functional group such as diethylaminoethyl (DEAE) and quaternary ammonium (Q) functional groups are employed for anion exchanger, while negatively charged functional groups such as sulfomethyl (S), carboxymethyl (CM), and sulfopropyl (SP) used for cation exchangers. Several ion exchange chromatography including DEAE Sepharose FF (GE Healthcare), DEAE-Sephadex (GE Healthcare), Q-Sepharose (GE Healthcare), and DEAE-cellulose (GE Healthcare) have been extensively used for purification of PBPs (Abalde et al. 1998; Liao et al. 2011; Sonani et al. 2016).

7.2.2.6 Aqueous Two-Phase System (ATPS)

The downstream processing required many steps for separation and purification of biomolecules that is very laborious and costly (Silva and Franco 2000). Therefore, demand has been increasing for cost-effective, economically feasible, and high yielding machinery for purification. The conventional chromatography technique of protein is not adequate for large-scale purification. The aqueous two-phase systems have overcome such limitations for optimum purification of proteins. Historically, aqueous two-phase system (ATPS) was discovered by Per-Åke Albertsson for purification of compounds (Albertsson 1986; Grilo et al. 2016). This is a kind of liquid–liquid fractionation technique and has a great potential for purification, separation, and enrichment of proteins, membranes, enzymes, nucleic acids, and biomolecules including viruses (Iqbal et al. 2016). This method has advantages over conventional purification system in terms of eco-friendly, low cost, continuous operation, etc. The ATPS system has been generated by mixing two polymers or polymer–salt together in water medium (Van Berlo et al. 1998). The affinity ligand-integrated ATPS may induce the rate of purification and recovery rate by hydrophobic interaction from polymer that is also termed as aqueous two-phase affinity partitioning (ATPAP) (Asenjo and Andrews 2011; Ruiz-Ruiz et al. 2012). The most common ATPS systems are formed by mixing of two polymers (polyethylene glycol (PEG) and dextran) or polymer–salt (polymer–phosphate or sulfate or citrate) (Iqbal et al. 2016). However, other types of ATPS such as ionic/liquid, short-chain alcohol, etc. also exist for separation and purification of other biomolecules (Hatti-Kaul 2001; Ruiz-Ruiz et al. 2012; Asenjo and Andrews 2012; Molino et al. 2013). The different molecular weight of PEGs polymer have low toxicity, cheap, low volatile in nature, thus it could be used in several ATPS for purification of biomolecules. The phase separation in ATPS should be optimized before purification in terms of salt composition, molecular weight, and concentration of polymer (Mazzola et al. 2008). The separation and purification for certain proteins are associated with partitioning pattern of PEG-PO_4 hydrophobicity exclusively and not dependent on molecular weight of proteins (Hachem et al. 1996). Further, Franco et al. (1996) established a mechanism of protein surface for hydrophobicity plays an important role in separation of biomolecules through ATPS. To reduce the demerit and better yield of PC, ATPS is formulated by mixing of PEG 4000 and potassium phosphate buffer (Zhao et al. 2014) through mechanism of droplet–droplet collision (Hatti-Kaul 2000). The separation of PC in ATPS is induced by fractional gravity and centrifugation (Chethana et al. 2015). Recently, an updated ATPS method has been introduced by interconnection of vortex fluidic device for enhancing phase recovery of native PC

(Luo et al. 2016). Thus, incorporation of ATPS with other device will be promising tools for higher recovery products.

7.3 Yield and Purity Index

The highest concentration of PBPs (PC and PE) is found in *Cyanophyceae* member *Arthrospira* sp. and *Rhodophyceae* member *Porphyridium* sp. (Roman et al. 2002; Sekar and Chandramohan 2008). Apart from the above species, *Nostoc* sp. and *Anabaena* sp. also have 10–17 % PC (*w/v*) (Moreno et al. 1995) while concentration in other cyanobacteria and red algae is altered in different environmental condition. The wide range of chromatography technique has been applied for separation and purification of PBPs microalgae crude extract (Sonani et al. 2016). Recently, numerous strategies of extraction and purification of PBPs from cyanobacteria have been discussed (Cuellar-Bermudez et al. 2015; Sonani et al. 2016). Here, we have shown few methods which applied for purification of PBPs from different cyanobacteria (Table 7.3).

A few methods have been commercialized that are scalable having low cost, few stages, high recovery, and yields for PBPs. Therefore, an utmost is a requirement to elaborate techniques for sustainable purification. Historically, Herrera et al. (1989) have followed downstream processing which includes ultrafiltration, adsorption on charcoal, and spray drying technique for purification of total cellular proteins and achieved good purity index (3.91) for PC with optimum yield (9 %) from total proteins. Recently, EBA chromatography integrated with anion exchanger (DEAE) has been developed to reduce the downstream processing, good recovery, and high yield of PBPs (Babu et al. 2006; Bermejo et al. 2006). Liu et al. (2005) found purity index 4.0–5.6 of R-PE extracted from *Polysiphonia urceolata* by using ion exchange column of DEAE–Sepharose Fast Flow column. The combination of anion exchange chromatography (DEAE–Sepharose) and gel filtration chromatography (Sephadex G-100) enhanced the purity index up to 5.06 and 5.34 for PC and APC, respectively (Zhang and Chen 1999). Similarly, anion exchange chromatography (DEAE-cellulose) was used to purify PBPs components from the cyanobacterium *Microcystis aeruginosa* (Padgett and Krogmann 1987). In addition, treatment of Rivanol for extraction of PC from *Spirulina fusiformis* was more advantageous in terms of higher yield and purity index (4.3) (Minkova et al. 2003). Benavides and Rito-Palomares (2006) found 2.9–3.2 purity index of B-PE extracted from red alga *Porphyridium cruentum* through ATPS chromatography. Patil et al. (2006) have obtained purity index 3.96 by adopting ATPS chromatography to purify PC from *Spirulina platensis*. The integration of ATPS and ion exchange chromatography has further increment 6.69 purity index of PC (Soni et al. 2008). However, combination of extraction protocol such as liquid nitrogen-induced cell breakage followed by lysozyme-triggered cell wall lysis was yielded 4.98 purity index of PC (Bhaskar et al. 2005). Apart from anion exchanger, HIC was also remarkable tools for hydrophobic-based separation (Niu et al. 2006), or it can be used in integration with ion exchange chromatography for efficient purification of R-PE from *Palmaria palmata* and *Polysiphonia urceolata* (Wang et al. 2002). The combination of

Table 7.3 List of certain relevant protocols employed for the purification of phycobiliproteins extracted from cyanobacteria and red algae

Species	Steps of PBPs purification	PC		PE		APC		References
		P^I	$Y^{\%}$	P^I	$Y^{\%}$	P^I	$Y^{\%}$	
Spirulina maxima	ATPS (PEG:phosphate); ultrafiltration	3.8	29	–	–	–	–	Rito-Palomares et al. (2001)
Spirulina fusiformis	Rivanol; ASP; Sephadex G-25	4.3	45	–	–	–	–	Minkova et al. (2003)
O. quadripunctulata	ASP; Sephadex G-150; Bio-Sil SEC; DEAE-cellulose	3.3	44	–	–	–	–	Soni et al. (2006)
Anabaena flos-aquae	Ultracentrifuge; HA column; HPLC (Vydac protein C-4)	4.7	–	–	–	–	–	Benedetti et al. (2006)
Spirulina platensis	ATPS(PEG:Phosphate); DEAE-Sephadex A-100	5.1	66	–	–	–	–	Patil et al. (2006)
Spirulina platensis	EBA; Q-Sepharose	3.6	8	–	–	–	–	Niu et al. (2007)
P. fragilis	ASP; methyl macro-prep HIC, HPLC (Bio-Sil SEC)	4.5	62	–	–	–	–	Soni et al. (2008)
Heterosiphonia japonica	ASP; Sepharose CL-4B; DEAE Sepharose fast flow	–	–	4.8	–	–	–	Sun et al. (2009)
Geitlerinema sp. A28DM	Ethodin; Sephadex G-100	–	–	–	–	3.2	62	Parmar et al. (2010)
Spirulina platensis	ASP; DEAE–Sepharose fast flow	5.5	67	–	–	5.1	80	Yan et al. (2011)
Pseudanabaena sp.	ASP; Sephadex G-150; DEAE- cellulose	–	–	6.8	47	–	–	Mishra et al. (2011)
Anabaena variabilis (CCC421)	ASP; DEAE-cellulose	–	–	4.9	62	–	–	Chakdar and Pabbi (2012)
Lyngbya sp. A09DM	Triton X-10; ASP; Sephadex G- 150; DEAE-cellulose	5.5	62	6.7	76	5.4	71	Sonani et al. (2014)
Nostoc sp. strain HKAR-11	ASP; Sephacryl S-100 HR, HiTrap phenyl fast flow	5.7	73	11	83	–	–	Kannaujiya et al. (2016)

ASP ammonium sulfate precipitation, *ATPS* aqueous two-phase system, *HPLC* high-performance liquid chromatography, *HA* hydroxylapatite, *PC* phycocyanin, *PE* phycoerythrin, *APC* allophycocyanin

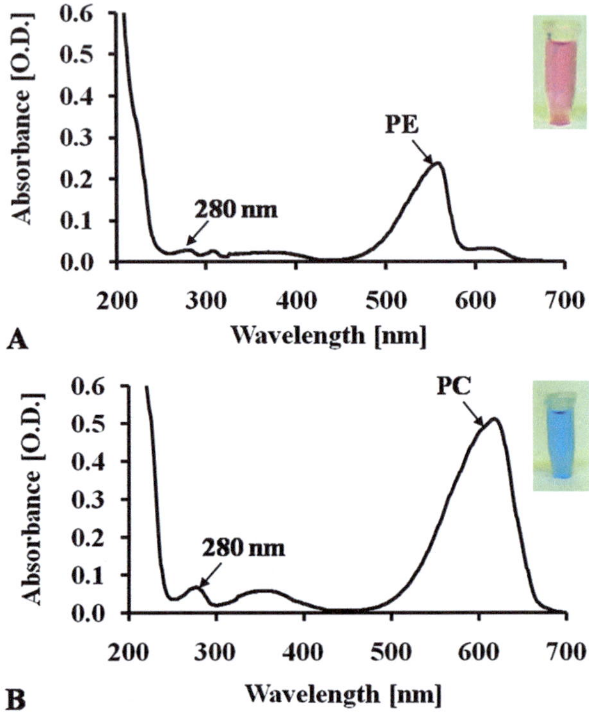

Fig. 7.4 Absorption spectrum of purified PE (**a**) and PC (**b**) after HiTrap Phenyl FF HIC column (flow rate of 60 ml h^{-1}). The elution of PE and PC was calculated by measuring the absorption at 280 nm (Adapted from Kannaujiya and Sinha 2016a)

hydrophobic and ion exchange chromatography shows mark increment of a purity index up to 4.85 of PC isolated from *Synechococcus* sp. IO9201 (Abalde et al. 1998). The lysozyme-treated extraction of PC from *Calothrix* sp. followed by Q-Sepharose column purification yielded 3.5 purity index (Santiago-Santos et al. 2004). Gupta and Sainis (2010) reported a single-step procedure for large-scale production of PC from *Anacystis nidulans* by charcoal–chitosan chromatography and obtained purity index 4.72 with high recovery. The hydrophobic-based single-step chromatography is more convenient for separation and purification of PC/PE (Kannaujiya and Sinha 2016a) (Fig. 7.4). The combination of gel filtration (Sephacryl S-100 HR) and HIC (HiTrap Phenyl FF) chromatography have achieved highest purity index (11.53) for PE (Kannaujiya and Sinha 2016a).

Isolation of PBPs from desiccation-tolerant cyanobacterium like *Lyngbya arboricola* is really tough due to the presence of extra sheath and dryness of cells. Tripathi et al. (2007) develop a novel strategy by using solvent-based precipitation and gel filtration and anion exchange chromatography for obtaining pure PE ($P^I > 5$) from the same cyanobacterium. Recently, ATPS-VFD chromatography has been developed for commercial purification (P^I 2.3) with high recovery yield (Y 78–93 %) from *Spirulina maxima* (Luo et al. 2016). Certain other chromatographic techniques such as polyethersulfone flat membrane (Marcati et al. 2014), Q-Sepharose (Senthilkumar et al. 2013), and affinity chromatography (Munier et al. 2015) further added advancement of purification of PC.

7.4 Market and Cost–Benefit Analysis

In current scenario, natural food color market was tremendously increasing worldwide. The data of market analysis of 2016 shows that food color business has reached US$ 1.3 billion with 6.8 % growth rate annually. The current growth rate is expected to reach US$1.77 billion up to 2021. According to new market analysis, America (N) and Europe (W) are two leading zones worldwide which have achieved 60 % market business collectively in 2015 and continuously upsurge owing to new food color pigment for further advancement. The extract from *Spirulina* sp. is the most demanding product and has achieved highest growth rate in 2016. On the basis of variety of pigment type, the total market can be categorized into certain raw products such as carotenoid, anthocyanin, phycobiliproteins (PBPs), carmine, betalains, and chlorophylls. The protein nature of PBPs plays a foremost role in market business with several valuable products. Currently, several companies have been established in pigment market and exploit commercial commodity with the novel bioproducts including PBPs. Cyanotech is developing large-scale microalgal products including PBPs for medical and biotechnology industries. There are many kinds of PBPs products developed such as conjugate R-PE, conjugate APC, stable cross-linked APC (XL-APC), etc. for diagnostic, flow cytometry and other biotechnology purposes (Sekar and Chandramohan 2008). Some other companies such as Dojindo, Flogen®, Martek Biosciences Corporation, Blue Biotech, Sanda King, EID Parry, Inner Mongolia Biomedical Eng., ProZyme, Pierce Biotechnology, Vectors Laboratories, AnaSpec Inc., Invitrogen Molecular Probes, Europa Bioproducts Ltd., and Chromaprobe Inc. have developed several PBP-based products for application in pharmaceutics, biotechnology, and biomedical sciences (Sekar and Chandramohan 2008) (Table 7.4).

Table 7.4 Name of companies and PBP-based biotechnological products

Company	Website address	Products
Cyanotech Corporation	http://www.phycobiliprotein.com	Conjugation kit: APC, B-PE, C-PC, R-PE
		Cross-linked: APC
ProZyme Inc.	http://www.prozyme.com/phycolink/pj-kits.html	Phycolink® conjugation kit: APC, B-PE,R-PE, C-PC,GT5APC, αGST (type 1)-R-PE, α-FLAG R-PE, goat α-human IgG R-PE
		PhycoPro™:APC, R-PC,Y-PE
Pierce Biotechnology Inc.	http://www.piercenet.com/products/browse.afm?	R-PE, PE monoclonal antibody (eBioPE-DLF), Zenon® R-PE, R-PE biotin XX, Alexa Fluor® 647-R-PE Streptavidin
Vectors Laboratories	http://www.vectorlabs.com/products.asp?catID-34	Conjugation kit: PE Avidin D, PE Streptavidin (SAPE)
Dojindo Molecular Technologies	http://www.dojindo.com/labeling/labelingkits.html	NH2 labeling kit: B-PE, R-PE, APC
		SH2 labeling kit: B-PE, R-PE, APC

(continued)

Table 7.4 (continued)

Company	Website address	Products
Flogen®	http://www.febico.com.tw/flogen/english/productflogen.htm	APC, B-PE, R-PE Cross-linked: CL-APC
AnaSpec Inc.	http://www.anaspec.com/products/product.category.asp?id=350	Protein labeling kit: AnaTag™ APC, AnaTag™ B-PE, AnaTag™ R-PE Cross-linked: CL-APC, B-PE, R-PE
Martek Bioscience Corporation	http://www.fluorescent.martek.com/technicaloverview	Pure products: R-PE, B-PE, APC SureLight® APC Sensilight™ dyes (stabilized phycobilisomes): PBXL-1, PBXL-3, P3L Sensilight™: P3L conjugates, PBXL-1 conjugates, PBXL-3 conjugates, APC conjugates
Invitrogen Molecular Probes	http://www.probes.invitrogen.com/servlets/pricelist?id=32093	Pure products: R-PE, B-PE,APC cross-linked: APXL, R-PE, Streptavidin conjugates: R-PE, B-PE, APC
Europa Bioproducts Ltd.	http://www.europabioproduct.com/catalogue/44	PhycoPro™: C-PC, Cross-linked CL-APC, B-PE,R-PE Phycolink®: -FLAG® R-PE conjugate, Streptavidin-R-PE
Chromaprobe Inc.	http://www.chromaprobe.com/products.html	R-PE, APC

Adapted and modified from Sekar and Chandramohan (2008)
APC Allophycocyanin, *B-PE Bangiales*-Phycoerythrin, *R-PE Rhodophyta*-Phycoerythrin, *R-PC Rhodophyta*-Phycocyanin, *C-PE Cyanophyta*-Phycocyanin

7.5 Conclusion

In the past century, the property of PBPs was widely studied to reveal structural and functional characteristics. In the past 20–30 years, distinguished application of PBPs has been established and commercialized in various fields of life sciences. However, only few cyanobacterial and red algal strains are involved in significant production of PBPs while numbers of strains are amenable for miniscule production. The basic and applied research is utmost need to upsurge more productive and fast-growing strains for better productivity. The purity of PBPs is playing a central role for quality of products. Currently, several techniques have been used for extraction and purification of PBPs. Indeed, none of the commercial method is developed which is applicable for extraction and purification of PBPs from any cyanobacterial sources with high yield and significant recovery. Thus, an urgent need is to develop a commercial technique for effective production and more utilization of PBPs to fulfill demand in mankind. The future depends on the growth and development of bioprocess engineering for manufacturing as well as improvement of commercial products.

References

Abalde J, Betancourt L, Torres E, Cid A, Barwell C (1998) Purification and characterization of phycocyanin from the marine cyanobacterium *Synechococcus* sp. IO9201. Plant Sci 136:109–120

Ajayan KV, Selvaraju M, Thirugnanamoorthy K (2012) Enrichment of chlorophyll and phycobiliproteins in *Spirulina platensis* by the use of reflector light and nitrogen sources: an in-vitro study. Biomass Bioenergy 47:436–441

Albertsson PÅ (1986) Partition of cell particles and macromolecules. Wiley Intersciences, Hoboken

Angelova MK, Dimov NP, Nachev ND, Tsvetkov AB (1997) Preparative hydrophobic interaction chromatography as effective stage in the purification of glucagon and insulin. Bulg Chem Commun 29:90–100

Anspach FB, Curbelo D, Hartmann R, Grake G, Deckwer W (1999) Expanded-bed chromatography in primary protein purification. J Chromatogr A 865:129–144

Asenjo JA, Andrews BA (2011) Aqueous two-phase systems for protein separation: a perspective. J Chromatogr A 1218:8826–8835

Asenjo JA, Andrews BA (2012) Aqueous two-phase systems for protein separation: phase separation and applications. J Chromatogr A 1238:1–10

Babu BR, Rastogi NK, Raghavarao MS (2006) Mass transfer in osmotic membrane distillation of phycocyanin colourant and sweet lime juice. J Membr Sci 212:58–69

Benavides J, Rito-Palomares M (2006) Simplified two-stage method to B-phycoerythrin recovery from *Porphyridium cruentum*. J Chromatogr B 844:39–44

Benedetti S, Benvenut F, Pagliarani S (2004) Antioxidant properties of a novel phycocyanin extract from the blue-green alga *Aphanizomenon flos-aquae*. Life Sci 75:2353–2362

Benedetti S, Rinalducci S, Benvenuti F, Francogli S, Pagliarani S, Giorgi L, Micheloni M, D'Amici GM, Zolla L, Canestrari F (2006) Purification and characterization of phycocyanin from blue-green alga *Aphanizomenon flos-aquae*. J Chromatogr B 833:12–18

Bermejo R, Ramos A (2012) Pilot scale recovery of phycocyanin from *Spirulina platensis* using expanded bed adsorption chromatography. Chromatographia 75:195–204

Bermejo R, Alvárez-Pez JM, Fernández AFG, Grima EM (2002) Recovery of pure B-phycoerythrin from the microalga *Porphyridium cruentum*. J Biotechnol 93:73–85

Bermejo R, Felipe MA, Talavera EM, Alvarez-Pez JM (2006) Expanded bed adsorption chromatography for recovery of phycocyanins from the microalga *Spirulina platensis*. Chromatographia 63:59–66

Bhaskar US, Gopalswamy G, Raghu R (2005) A simple method for efficient extraction and purification of C-Phycocyanin from *Spirulina platensis* Geitler. Indian J Exp Biol 43:277–279

Boussiba S, Richmond AE (1979) Isolation and characterization of phycocyanins from the blue-green alga *Spirulina platensis*. Arch Microbiol 120:155–159

Boussiba S, Richmond A (1980) C-Phycocyanin as a storage in the blue-green alga *Spirulina platensis*. Arch Microbiol 125:143–147

Bryant DA, Glazer AN, Eiserling FA (1976) Characterization and structural properties of the major biliproteins of *Anabaena* sp. Arch Microbiol 110:61–75

Bryant DA, Guglielmi G, Tandeau de Marsac N, Castlets AM, Cohen-Bazire G (1979) The structure of cyanobacterial phycobilisomes: a model. Arch Microbiol 123:113–127

Chakdar H, Pabbi S (2012) Extraction and purification of phycoerythrin from *Anabaena variabilis* (CCC421). Phykos 42:25–31

Chase HA, Draeger NM (1992) Expanded bed adsorption of proteins using ion exchangers. Sep Sci Technol 27(14):2021–2039

Chemat F, Vian MA, Cravotto G (2012) Green extraction of natural products: concept and principles. Int J Mol Sci 13:8615–8627

Chethana S, Nayak C, Madhusudhan MC, Raghavarao KSMS (2015) Single step aqueous two-phase extraction for downstream processing of C-phycocyanin from *Spirulina platensis*. J Food Sci Technol 52:2415–2421

Cohen Z (1997) The chemicals of *Spirulina*. In: Vonshak A (ed) Spirulina platensis (Arthrospira): physiology, cell biology and biotechnology. Taylor and Francis, London, pp 175–204

Cuellar-Bermudez SP, Aguilar-Hernandez I, Cardenas-Chavez DL, Ornelas-Soto N, Romero-Ogawa MA, Parra-Saldivar R (2015) Extraction and purification of high-value metabolites from microalgae: essential lipids, astaxanthin and phycobiliproteins. Microb Biotechnol 8:190–209

Esquivel-Hernández DA, López VH, Rodríguez-Rodríguez J, Alemán-Nava GS, Cuéllar-Bermúdez SP, Rostro-Alanis M, Parra-Saldívar R (2016) Supercritical carbon dioxide and microwave-assisted extraction of functional lipophilic compounds from *Arthrospira platensis*. Int J Mol Sci 17:658

Fausnaugh JL, Regnier FE (1986) Solute and mobile phase contributions to retention in hydrophobic interaction chromatography of proteins. J Chromatogr 359:131–146

Franco T, Andrews A, Asenjo J (1996) Use of chemically modified proteins to study the effect of a single protein property on partitioning in aqueous two phase systems: effect of surface charge. Biotechnol Bioeng 49:309–315

Garcia FAP, Pires EMV (1993) Chromatography. In: Kennedy JF, Cabral JMS (eds) Recovery processes for biological materials. Wiley, London, pp 415–451

Glauser M, Bryant DA, Frank G, Wehrli E, Rusconi SS, Sidler W, Zuber H (1992) Phycobilisome structure in the cyanobacteria *Mastigocladus laminosus* and *Anabaena* sp. PCC 7120. Eur J Biochem 205:907–915

Glazer AN (1984) Phycobilisome: a macromolecular complex optimized for light energy transfer. Biochim Biophys Acta 768:29–51

Glazer AN (1985) Light harvesting by phycobilisomes. Annu Rev Biophys Chem 14:47–77

Gonzalez Y, Ibarra N, Gomez H, Gonzalez M, Dorta L, Padilla S, Valdes R (2003) Expanded bed adsorption processing of mammalian cell culture fluid: comparison with packed bed affinity chromatography. J Chromatogr B 784:183–187

Green AA, Hughes WL (1955) Protein solubility on the basis of solubility in aqueous solutions of salts and organic solvents. Methods Enzymol 1:67–90

Grilo AL, Raquel Aires-Barros M, Azevedo AM (2016) Partitioning in aqueous two-phase systems: fundamentals, applications and trends. Sep Purif Rev 45:68–80

Gupta A, Sainis JK (2010) Isolation of C-phycocyanin from *Synecococcus* sp., (*Anacystis nidulans* BD1). J Appl Phycol 22:231–233

Hachem F, Andrews B, Asenjo J (1996) Hydrophobic partitioning of proteins in aqueous two-phase systems. Enzym Microb Technol 19:507–517

Hatti-Kaul R (2000) Aqueous two-phase systems. Methods and protocols. Humana Press, Totowa

Hatti-Kaul R (2001) Aqueous two-phase systems. Mol Biotechnol 19:269–277

Herrera A, Boussiba S, Napoleone V, Hohlberg A (1989) Recovery of C-phycocyanin from the cyanobacterium *Spirulina maxima*. J Appl Phycol 1:325–331

Hjertén S (1973) Some general aspects of hydrophobic interaction chromatography. J Chromatogr 87:325–331

Horváth H, Kovács AW, Riddick C, Présing M (2013) Extraction methods for phycocyanin determination in freshwater filamentous cyanobacteria and their application in a shallow lake. Eur J Phycol 48:278–286

Iqbal M, Tao Y, Xie S, Zhu Y, Chen D, Wang X, Huang L, Peng D, Sattar A, Shabbir MAB, Hussain HI, Ahmed S, Yuan Z (2016) Aqueous twophase system (ATPS): an overview and advances in its applications. Biol Proced Online 18:18

Johnson EM, Kumar K, Das D (2014) Physicochemical parameters optimization, and purification of phycobiliproteins from the isolated *Nostoc* sp. Bioresour Technol 166:541–547

Juin C, Chérouvrier JR, Thiéry V, Gagez AL, Bérard JB, Joguet N, Kaas R, Cadoret JP, Picot L (2015) Microwave-assisted extraction of phycobiliproteins from *Porphyridium purpureum*. Appl Biochem Biotechnol 175:1–15

Kannaujiya VK, Sinha RP (2015) Impacts of varying light regimes on phycobiliproteins of *Nostoc* sp. HKAR-2 and *Nostoc* sp. HKAR-11 isolated from diverse habitats. Protoplasma 252(6):1551–1561

Kannaujiya VK, Sinha RP (2016a) An efficient method for separation and purification of phycobiliproteins from rice-field cyanobacterium *Nostoc* sp. strain HKAR-11. Chromatographia 79:335–343

Kannaujiya VK, Sinha RP (2016b) Thermokinetic stability of phycocyanin and phycoerythrin in food grade preservatives. J Appl Phycol 28:1063–1070

Kannaujiya VK, Rahman A, Adinath SS, Sinha RP (2016) Structural and functional dynamics of tyrosine amino acid in phycocyanin of hot-spring cyanobacteria: a possible pathway for internal energy transfer. Gene Rep 5:83–91

Kilpatrick KA (1985) The development of a method for measure marine cyanobacterial phycoerythrin extracted in solvents. Dissertation, Texas AM University

Li P, Gomes PF, Loureiro JM, Rodrigues AE (2014) Proteins separation and purification by expanded bed adsorption and simulated moving bed technology. In: Subramanian G (ed) Continuous processing in pharmaceutical manufacturing. Wiley-VCH Verlag GmbH & Co. KGaA, Weinheim

Liao X, Zhang B, Wang X et al (2011) Purification of c-phycocyanin from *Spirulina platensis* by single-step ion-exchange chromatography. Chromatographia 73(3):291–296

Liu L, Chen X, Zhang X, Zhang Y, Zhou B (2005) One-step chromatography method for efficient separation and purification of R-phycoerythrin from *Polysiphonia urceolata*. J Biotechnol 116:91–100

Luo X, Smith P, Raston CL, Zhang W (2016) Vortex fluidic device-intensified aqueous two phase extraction of c-phycocyanin from *Spirulina maxima*. ACS Sustain Chem Eng 4(7):3905–3911

Marcati A, Ursu AV, Laroche C, Soanen N, Marchal L, Jubeau S, Djelveh G, Michaud P (2014) Extraction and fractionation of polysaccharides and B-phycoerythrin from the microalga *Porphyridium cruentum* by membrane technology. Algal Res 5:258–263

Mazzola PG, Lopes AM, Hasmann FA, Jozala AF, Penna TC, Magalhaes PO, Rangel-Yagui CO, Pessoa A Jr (2008) Liquid-liquid extraction of biomolecules: an overview and update of the main techniques. J Chem Technol Biotechnol 83:143–157

Minkova KM, Tchernov AA, Tchorbadjieva MI, Fournadjieva ST, Antova RE, Busheva MC (2003) Purification of C-phycocyanin from *Spirulina (Arthrospira) fusiformis*. J Biotechnol 102:55–59

Mishra SK, Shrivastav A, Mishra S (2011) Preparation of highly purified C-phycoerythrin from marine cyanobacterium *Pseudanabaena* sp. Protein Expr Purif 80:234–238

Molino JVD, Marques V, de Araújo D, Júnior AP, Mazzola PG, Gatti MSV (2013) Different types of aqueous two-phase systems for biomolecule and bioparticle extraction and purification. Biotechnol Prog 29:1343–1353

Moraes C, Sala L, Cerveira GP, Kalil SJ (2011a) C-phycocyanin extraction from Spirulina platensis wet biomass. Braz J Chem Eng 28(1):45–49

Moraes CC, da Costa Ores J, Costa JAV, Kalil SJ (2011b) Recovery of c-phycocyanin in the presence of cells using expanded bed IEC. Chromatographia 74(3):307–312

Moreno J, Rodriquez H, Vargas MA, Rivas J, Guerrero MG (1995) Nitrogen fixing cyanobacteria as a source of phycobiliproteins pigments-composition and growth performance of ten filamentous herterocystous strains. J Appl Phycol 7:17–23

Munier M, Moranc M, Dumay J, Jaouen P, Fleurence J (2015) One-step purification of R-phycoerythrin from the red edible seaweed *Grateloupia turuturu*. J Chromatogr B 992:23–29

Muramatsu M, Hihara Y (2012) Acclimation to high-light conditions in cyanobacteria: from gene expression to physiological responses. J Plant Res 125:11–39

Niu JF, Wang GC, Tseng CK (2006) Method for large-scale isolation and purification of R-phycoerythrin from red alga *Polysiphonia urceolata* Grev. Protein Expr Purif 49:23–31

Niu J-F, Wang G-C, Lin X-Z, Zhou B-C (2007) Large-scale recovery of C-phycocyanin from *Spirulina platensis* using expanded bed adsorption chromatography. J Chromatogr B Analyt Technol Biomed Life Sci 850:267–276

Padgett MP, Krogmann DW (1987) Large scale preparation of pure phycobiliproteins. Photosynth Res 11:225–235

Parmar A, Singh NK, Madamwar D (2010) Allophycocyanin from a local isolate *Geitlerinema* sp. A28DM (cyanobacteria): a simple and efficient purification process. J Phycol 46:285–289

Parmar A, Singh NK, Kaushal A, Sonawala S, Madamwar D (2011) Purification, characterization and comparison of phycoerythrins from three different marine cyanobacterial cultures. Bioresour Technol 102:1795–1802

Patil G, Chethana S, Sridevi AS, Raghavarao KSMS (2006) Method to obtain C-phycocyanin of high purity. J Chromatogr A 1127:76–81

Pierce JJ, Fischer EJ, Smith MP, Turner C, Keshavarz-Moore E, Dunnill P (1999) Purification of a periplasmic enzyme by a stirred adsorbent, and by expanded and packed beds. Bioprocess Eng 20:449–457

Queiroz JA, Garcia FAP, Cabral JMS (1999) Purification of *Chromobacterium viscosum* lipase by hydrophobic interaction chromatography. Isolat Purif 3:1–7

Queiroz JA, Tomaz CT, Cabral JMS (2001) Hydrophobic interaction chromatography of proteins. J Biotechnol 87:143–159

Ramos A, Acién FG, Fernández-Sevilla JM, González CV, Bermejo R (2010) Large-scale isolation and purification of C-phycocyanin from the cyanobacteria Anabaena marina using expanded bed adsorption chromatography. J Chem Technol Biotechnol 85(6):783–792

Ramos A, Acién FG, Fernández-Sevilla JM, González CV, Bermejo R (2011) Development of a process for large-scale purification of C-phycocyanin from *Synechocystis aquatilis* using expanded bed adsorption chromatography. J Chromatogr B 879:511–519

Ranjitha K, Kaushik BD (2005) Purification of phycobiliproteins from *Nostoc muscorum*. J Sci Ind Res 64:372–375

Reis A, Mendes A, Lobo-Fernandes H, Empis JA, Novais JM (1998) Production, extraction and purification of phycobiliproteins from *Nostoc* sp. Bioresour Technol 66:181–187

Richa RRP, Kumari S, Singh KL, Kannaujiya VK, Singh G, Kesheri M, Sinha RP (2011) Biotechnological potential of mycosporine-like amino acids and phycobiliproteins of cyanobacterial origin. Biotechnol Bioinformatics Bioeng 1:159–171

Rito-Palomares M, Nunez L, Amador D (2001) Practical application of aqueous two-phase systems for the development of a prototype process for C-phycocyanin recovery from *Spirulina maxima*. J Chem Technol Biotechnol 76:1273–1280

Rodríguez H, Rivas J, Guerrero MG, Manuel Losada M (1991) Enhancement of phycobiliprotein production in nitrogen-fixing cyanobacteria. J Biotechnol 20:263–270

Roe S (1989) Purification based on hydrophobicity. In: Harris ELV, Angal S (eds) Protein purification methods: a practical approach. IRL Press, Oxford, pp 221–232

Roman RB, Alvárez-Pez JM, Fernández FGA, Grima EM (2002) Recovery of pure B- Phycoerythrin from the microalga *Porphyridium cruentum*. J Biotechnol 93:73–85

Ruiz-Ruiz F, Benavides J, Aguilar O, Rito-Palomares M (2012) Aqueous two-phase affinity partitioning systems: current applications and trends. J Chromatogr A 1244:1–13

Rupley JA, Gratton E, Careri G (1983) Water and globular proteins. Trends Biochem Sci 8:18–22

Sala L, Figueira FS, Cerveira GP, Moraes CC, Kalil SJ (2014) Kinetics and adsorption isotherm of C-phycocyanin from *Spirulina platensis* on ion-exchange resins. Braz J Chem Eng 31:1013–1022

Santiago-Santos MC, Ponce-Noyola T, Olvera-Ramírez R, Ortega-López J, Cañizares-Villanueva RO (2004) Extraction and purification of phycocyanin from *Calothrix* sp. Process Biochem 39:2047–2052

Sarada R, Pillai MG, Ravishankar GA (1999) Phycocyanin from *Spirulina* sp: influence of processing of biomass on phycocyanin yield, analysis of efficacy of extraction methods and stability studies on phycocyanin. Process Biochem 34(8):795–801

Sekar S, Chandramohan M (2008) Phycobiliproteins as a commodity: trends in applied research, patents and commercialization. J Appl Physiol 20:113–136

Senthilkumar N, Suresh V, Thangam R, Kurinjimalar C, Kavitha G, Murugan P, Kannan S, Rengasamy R (2013) Isolation and characterization of macromolecular protein R-Phycoerythrin from *Portieria hornemannii*. Int J Biol Macromol 55:150–160

Sidler WA (1994) Phycobilisome and phycobiliprotein structures. In: Bryant DA (ed) The molecular biology of cyanobacteria. Kluwer Academic Publication, Netherlands, pp 139–216

Silman O, Wilkinson J, Burshteyn A, Roth P, Ledis S (1999) Fluorescent neoglycoproteins: antibody-aminodextran-phycobiliprotein conjugates. Bioconjug Chem 10:1090–1106

Silva M, Franco T (2000) Liquid-liquid extraction of biomolecules in downstream processing-a review paper. Braz J Chem Eng 17:1–17

Simpson RJ (2010) Disruption of cultured cells by nitrogen cavitation. Cold Spring Harb Protoc 11:pdb.prot5513

Singh SP, Montgomery BL (2013) Salinity impacts photosynthetic pigmentation and cellular morphology changes by distinct mechanisms in *Fremyella diplosiphon*. Biochem Biophys Res Commun 433:84–89

Singh NK, Parmar A, Madamwar D (2009) Optimization of medium components for increased production of C-phycocyanin from *Phormidium ceylanicum* and its purification by single step process. Bioresour Technol 100:1663–1669

Sinha RP, Lebert M, Kumar A, Kumar HD, Häder D-P (1995) Spectroscopic and biochemical analyses of UV effects on phycobiliproteins of *Anabaena* sp. and *Nostoc carmium*. Bot Acta 108:87–92

Sonani RR, Singh NK, Kumar J, Thakar D, Madamwar D (2014) Concurrent purification and antioxidant activity of phycobiliproteins from *Lyngbya* sp. A09DM: an antioxidant and anti-aging potential of phycoerythrin in *Caenorhabditis elegans*. Process Biochem 49:1757–1766

Sonani RR, Rastogi RP, Patel R, Madamwar D (2016) Recent advances in production, purification and applications of phycobiliproteins. World J Biol Chem 26:100–109

Soni B, Kalavadia B, Trivedi U, Madamwar D (2006) Extraction, purification and characterization of phycocyanin from *Oscillatoria quadripunctulata* isolated from the rocky shores of Bet-Dwarka, Gujarat, India. Process Biochem 41:2017–2023

Soni B, Trivedi U, Madamwar D (2008) A novel method of single step hydrophobic interaction chromatography for the purification of phycocyanin from *Phormidium fragile* and its characterization for antioxidant property. Bioresour Technol 99:188–194

Stewart DE, Farmer FH (1984) Extraction, identification and quantitation of phycobiliproteins pigments from phototropic biomass. Limnol Oceanogr 29:392–397

Sudhakar MP, Jagatheesan A, Perumal K, Arunkumar K (2015) Methods of phycobiliprotein extraction from *Gracilaria crassa* and its applications in food colourants. Algal Res 8:115–120

Sun L, Wang S, Gong X, Zhao M, Fu X, Wanga L (2009) Isolation, purification and characteristics of R-phycoerythrin from a marine macroalga *Heterosiphonia japonica*. Protein Expr Purif 64:146–154

Tandeau de Marsac N, Cohen-Bazire G (1977) Molecular composition of cyanobacterial phycobilisomes. Proc Natl Acad Sci U S A 74:1635–1639

Thomas J-C, Passaquet C (1999) Characterization of a phycoerythrin without α-subunits from a unicellular red alga. J Biol Chem 274:2472–2482

Tripathi SN, Kapoor S, Shrivastava A (2007) Extraction and purification of an unusual phycoerythrin in a terrestrial desiccation tolerant cyanobacterium *Lyngbya arboricola*. J Appl Phycol 19:441–447

Tsai PL, Chen CH, Huang CJ, Chou CM, Chang GD (2004) Purification and cloning of an endogenous protein inhibitor of carp nephrosin, an astacin metalloproteinase. J Biol Chem 279(12):11146–11155

Van Berlo M, Luyben KCA, van der Wielen LA (1998) Poly (ethylene glycol)-salt aqueous two-phase systems with easily recyclable volatile salts. J Chromatogr B 711:61–68

Vernet M, Mitchell EG, Holm-Hansen O (1990) Adaptation of *Synechococcus* in situ determined by variability in intracellular phycoerythrin-543 at a coastal station of the Southern California coast, USA. Mar Ecol Prog Ser 63:9–16

Viskari PJ, Colyer CC (2003) Rapid extraction of phycobiliproteins from cultured cyanobacteria samples. Anal Biochem 319:263–271

Wang GC, Sun HB, Fan X, Tseng CK (2002) Functional ingredients from algae for foods and nutraceuticals. Acta Bot Sin 44:541–546

Wang L, Qu Y, Fu X, Zhao M, Wang S, Sun L, Rozhkova EA (2014) Isolation, purification and properties of an R-phycocyanin from the phycobilisomes of a marine red macroalga *Polysiphonia urceolata*. PLoS ONE 9:e87833

Wiltshire KH, Boersma M, Moller A et al (2000) Extraction of pigments and fatty acids from the green alga *Scenadesmus oblique* (cholorophycae). Aquat Ecol 34:119–126

Wingfield PT (2001) Protein precipitation using ammonium sulfate. Curr Protoc Protein Sci:1–10

Wood JL, Miller CD, Sims RC, Takemoto JY (2015) Biomass and phycocyanin production from cyanobacteria dominated biofilm reactors cultured using oilfield and natural gas extraction produced water. Algal Res 11:165–168

Wyman M (1992) An in vivo method for the estimation of phycoerythrin concentration in marine cyanobacteria (*Synechococcus* spp.) Limnol Oceanogr 37:1300–1306

Yan SG, Zhu LP, Su HN, Zhang XY, Chen XL, Zhou BC, Zhang YZ (2011) Single-step chromatography for simultaneous purification of C-phycocyanin and allophycocyanin with high purity and recovery from *Spirulina* (*Arthrospira*) *platensis*. J Appl Phycol 23:1–6

Zhang YM, Chen F (1999) A simple method for efficient separation and purification of C-phycocyanin and allophycocyanin from *Spirulina platensis*. Biotechnol Tech 13:601–603

Zhao L, Peng Y, Gao J, Cai W (2014) Bioprocess intensification: an aqueous two-phase process for the purification of C-phycocyanin from dry *Spirulina platensis*. Eur Food Res Technol 238:451–457

Zhao M, Sun L, Sun S, Gong X, Fu X, Chen M (2015) Phycoerythrins in phycobilisomes from the marine red alga *Polysiphonia urceolata*. Int J Biol Macromol 73:58–64

Food and Biotechnological Applications 8

8.1 Introduction

The natural colors of cyanobacteria, red algae, and cryptomonads originate from brilliantly colored light-harvesting proteins called phycobiliproteins (PBPs) (Glazer 1981). There are wide varieties of accessory light-harvesting complexes in habitats of specific micro-/macroalgae to harvest a broad spectrum of photonic energy for enhancement of photosynthesis in various environmental conditions (Grossman et al. 1995; Kannaujiya and Sinha 2015). PBPs are made up of peripheral rod and core biliproteins, which consists of phycocyanin (PC), phycoerythrin (PE), and allophycocyanin (APC) united by linker polypeptides (Fig. 8.1).

Commonly, PBPs constitute 24% of the dry mass of the total cellular proteins (Glazer 1985; Grossman et al. 1993; Cai et al. 2001). However, PBP concentrations can be enhanced up to 40% under low light intensity (Glazer 1994). Interestingly, in the past few years, PBPs have played a major role in the development of commercial food colors, nutraceuticals, and other biotechnological products. Recently, PBPs have gained further potential roles in pharmaceuticals, cosmetics, and fluorescent agents (Richa et al. 2011). Current utilization of synthetic colors for coloration of food materials involves toxic elements that are potentially harmful to the human body and play important roles in causing cancerous disease. They are broadly used in making food products, beverages, dairy products, ice cream, candy, etc. Besides their properties as natural colorants, PBPs have unique fluorescent properties that are enormously useful for pharmaceutical, biomedical, molecular biology, and therapeutic applications (Eriksen 2008) (Fig. 8.2).

Currently, a few organisms are exploited for production of PBPs, including *Spirulina* sp. and *Porphyridium* sp. (Roman et al. 2002). There are hundreds of patents involving PBPs in various fields, such as 55 patents for production of PBPs, 236 for fluorescence applications, and 30 relating to the clinical importance of PBPs (Sekar and Chandramohan 2008). The aim of this chapter is to define the broad and prosperous applications of PBPs in the fields of biotechnology, pharmacology, and food applications, for commercial utilization.

V.K. Kannaujiya et al., *Phycobiliproteins: Recent Developments and Future Applications*, https://doi.org/10.1007/978-981-10-6460-9_8

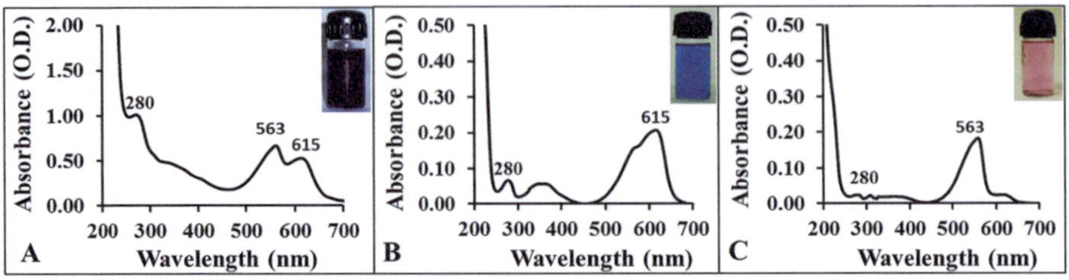

Fig. 8.1 Absorbance bands of (**a**) crude extracts, (**b**) phycocyanin (PC), and (**c**) phycoerythrin (PE) purified from *Nostoc* sp. strain HKAR-2 (Adapted from Kannaujiya and Sinha 2016)

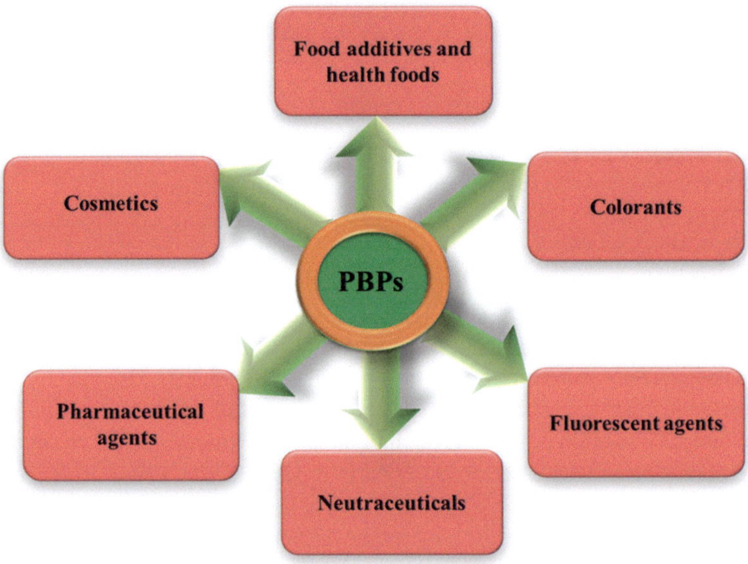

Fig. 8.2 Commercial applications of phycobiliproteins (PBPs)

8.2 Nutraceuticals and Natural Food Colorants

The US Food and Drug Administration (FDA) regulates the identification and manipulation of nutritional value in the design of effective food products for promotion of health and immunization against common diseases. Now, a branch of science and technology has developed many functional foods that promise remarkable health benefits for humans and improved quality of life. Seaweeds are very significant foods, containing large amounts of polysaccharides, minerals, proteins, lipids, polyphenols, and vitamins, with indispensable properties (Arasaki and Arasaki 1983; Kumar et al. 2008). In the industrial sector, many companies are producing food supplements and nutraceuticals from *Spirulina* sp. and utilizing them in medicines. DIC International (formerly Dainippon Ink and Chemicals) is one of the biggest companies involved in food product production from *Spirulina* sp.; its estimated gross production is over 300 t per annum and is increasing over time (Lee 1997; DIC 2009). Earthrise® Nutritionals, which is part of DIC, owns the world's largest *Spirulina* farm, in California, USA, which covers over 444,000 m^2. The *Spirulina*-based tablets it produces are marketed in more than 20 countries (Earthrise Nutritionals LLC 2004). Cyanotech Corporation, situated on the island of Hawaii (in the Kona district), produces *Spirulina*-based products (Cyanotech Corporation 2007) used in traditional food products—including biscuits, breakfast cereals, desserts, and pasta—which are largely consumed in different European countries instead of synthetic nutraceuticals. Commercially, PBPs are produced from *Spirulina* sp., *Porphyridium* sp., and *Rhodella* sp. for preparation of food additives and dyes (Singh et al. 2005; Spolaore et al. 2006). However, blue food coloring (PC) is not considered to be commercial food grade in European countries (Prasanna et al. 2007; Eriksen 2008). A few studies have been concerned about the application of PC as a food dye related to color preservatives (Jespersen et al. 2005; Mishra et al. 2010, 2012). A study has also been done on the rheological properties of PC (Batista et al. 2006). Currently, PBPs are used as dye for developing colored food products such as dairy products and confectionary (Sekar and Chandramohan 2008). Certain companies in Tokyo, such as DKSH Japan and DIC Corporation, have marketed PC as a natural dye (Houghton 1996). The yield of PE from red algae such as *Porphyridium* sp. is approximately 200 mg/L, and it is used in colored confectionary and gelatin desserts (Dufosse et al. 2005). With regard to its dye properties, PE has bright yellow fluorescence, which plays an important role in special effects during dark conditions under exposure to ultraviolet (UV) radiation. A wide range of food products are prepared that glow under UV radiation. A range of tested foods—such as lollipops, soft drinks, dry sugar-drop candies, and alcoholic beverages—fluoresce under exposure to UV radiation at an acidic pH of 5–6. Similarly, the fluorescent nature of PC has been preserved in 30% alcohol for color stability (Dufosse et al. 2005). The hot-spring-isolated cyanobacterium *Nostoc* sp. strain HKAR-2 produces PC which is more stable to heat and sensitive to pH and light (Kannaujiya and Sinha 2016). Light and pH stability are important for reducing the degradation of colored products such as beverages. Jelly gums and candy productions by using PC colorant have increased in confectionary. The appearance of the blue color of PC is denatured in hard candy; however, the blue color in jelly gum can be sustained by the presence of a gelatin matrix (Jespersen et al. 2005). Jelly gum centers become discolored when exposed to intense solar radiation for 24 h. PE requires use of long-term preservatives to keep it safe and unspoiled, and it has high sensitivity to light and the oxygen activity of food products (Mishra et al. 2010). Nonpurified forms of PC isolated from a crude extract of *Arthrospira platensis* are widely accepted in the form of antioxidant health foods (Estrada et al. 2001; Bermejo et al. 2008). The crude extract and whole cyanobacteria also contain various kind of metabolites, consumption of which has been credited with positive cholesterol-lowering, anti-inflammatory, anticancer, and antiviral effects (Jensen et al. 2001; Singh et al. 2005). Recently, there has been more focus on the use of PC

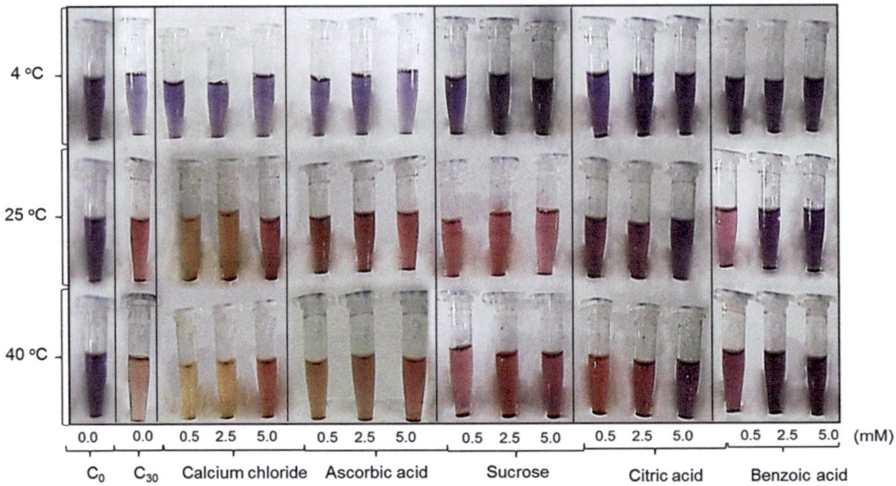

Fig. 8.3 Effects of various concentrations of different preservatives on blue food color (phycocyanin [PC]) and red food color (phycoerythrin [PE]) stored at 4, 25, and 40 °C. C_0 control at day 0, C_{30} control at day 30 (Adapted from Kannaujiya and Sinha 2016)

as a nutraceutical, which is obtained from dried biomass of *A. platensis*. The production of PC has increased globally to a total market value of up to US$10–50 million/year and a value-added colored food product value of around US$450/kg (Bhaskar et al. 2005; Spolaore et al. 2006; Leema et al. 2010). Proper application of any natural colorant in food requires detailed knowledge about the stability of the colorant in order to optimize large-scale production of PBPs for this application. The blue color of PC is added to soft beverages (e.g., Pepsi® and Bacardi Breezer®), which retain the color for about 1 month at optimum temperature. The natural blue colorant PC is very significantly used in dry confectionary with involvement of sugar. This makes it very stable for flower cake decorations, which easily retain the color for a year. PBPs are very sensitive to light, temperature, and the pH of the medium, and they are easily degraded by alterations in these factors. Maintenance of food products for a long time and applications in various fields of science are critical tasks for the biotechnology industry. Food-grade preservatives play a significant role in enhancement of the stability of food products under storage conditions (Brul and Coote 1999). Addition of some food-grade preservatives such as sucrose, sodium chloride, and citric acid has been reported for long-term heat stability of PC (Chaiklahan et al. 2012). Benzoic acid is a good preservative agent for expanding the life of PBPs in biological food products stored at 4, 25, and 40 °C (Kannaujiya and Sinha 2016) (Fig. 8.3), possibly because benzoic acid acts as an antioxidant and its antimicrobial activity inhibits growth of microbes [(Dong and Wang 2006), (Lino and Pena 2010)].

8.2.1 Legislation on Food Supplements

Micro-/macroalgal food products come under consumer food safety regulations to ensure the quality of food products. In France, seaweed products have been authorized for consumption by humans. Interest in consumption and the functional role of organic food products is increasing worldwide. However, currently no definition has been published for categorization of food products. In Spain, contamination of food products is not covered by the relevant legislation. The food industry is flourishing in Japan, the USA, Ireland, India, Europe and Asia–Pacific countries. Market analysis shows that the combined revenue for food products in the USA, Western Europe and Asia–Pacific countries is growing continuously, with an average of 5.7% annual growth rate (Holdt and Kraan 2011). In the USA, food products are regulated by the FDA. A similar authority exists in Japan. In 2003, Ireland established certain rules for utilization of mineral nutrients as food supplements, and these are now regulated by the Irish Food Safety Authority (FSAI). Recently, the European Food Safety Authority (EFSA) has introduced legislation and safety assessment rules for authorization and introduction of novel foods and ingredients.

8.3 Pharmaceuticals and Fluorescence Probes

8.3.1 Pharmaceutical Agents

In the last few decades, the green approach for generation of antibiotics, pharmaceutical products, and fluorescent tags has received interest for industrial production (Sekar and Chandramohan 2008). PC has a great advantage over the pharmacological properties of other compounds in term of antioxidant, anticancer, anti-inflammatory, anti-aging, hepatoprotective, and neuroprotective activities useful for human welfare (Table 8.1). Evaluation of the antioxidant properties of PC demonstrates scavenging of hydroxyl, alkoxyl, peroxide, superoxide, and nitrogen-containing radicals, with a probable role in inhibition of lipid peroxidation. PC sourced from *Aphanizomenon flos-aquae* exhibits strong antioxidative properties (Bhat and Madyastha 2000; Romay et al. 2003) and clearly demonstrates the ability to protect against several types of oxidative damage (Benedetti et al. 2004). Similarly, PC plays a significant role in inhibition of oxidative hemolysis of red blood cells triggered by peroxyl radicals. PC has been shown to significantly reduce lipid peroxidation in plasma samples exposed to pro-oxidant synthetic chemicals such as cupric chloride. These PC-induced antioxidative phenomena could help to prevent many oxidation-induced pathological disorders. PC is able to inhibit development of edema and excessive release of histamine, prostaglandin E2 (PGE2), myeloperoxide (MPO), and leukotriene B4 (LTB4) in tissue during inflammation (Richa et al. 2011).

In addition, PC from *Spirulina platensis* shows significant inhibitory effects on the growth and development of cancer in terms of dose dependency and time

Table 8.1 Potential pharmaceutical properties and modes of action of phycobiliproteins (PBPs)

Potential pharmaceutical property	Mode of action	Experimental system	References
Antioxidant	Radical-scavenging activities	Mouse	Sekar and Chandramohan (2008) and Eriksen (2008)
Anti-inflammatory	Inhibition of glucose oxidase, edema, colitis, and oxygen radicals	Mouse	Romay et al. (1998), González et al. (1999), and Sekar and Chandramohan (2008)
Anti-atherosclerosis	Inhibition of atherosclerosis development, enhancement of antioxidant effect	Hamster	Riss et al. (2007)
Neuroprotective	Inhibition of neural damage process by kainic acid, antioxidant properties	Rat	Rimbau et al. (1999)
Inhibition of stone formation and lipid peroxidation	Inhibition of stone formation by oxalic acid, inhibition of lipid peroxidation	Rat	Farooq et al. (2004)
Inhibition of drug toxicity	Reduction of cardiotoxicity of doxorubicin, scavenging of oxygen radicals	Rat	Kahn et al. (2006)
Inhibition of nitric oxide synthesis	Inhibition of nitric oxide and nitrite synthesis	Macrophage cells	Cherng et al. (2007)
Anticancer	Hepatocellular apoptotic inducement, inhibition of leukemia cells	Hepatocellular cells, leukemia cells	Roy et al. (2007), Liu et al. (2000), and Subhashini et al. (2004)

Adapted and modified from Eriksen (2008). For more details, see reference

duration in cell lines of human leukemia K562 (Liu et al. 2000). Cholesterol levels are also reduced by PC, which may help to inhibit development of heart disease (Nagaoka et al. 2005). PC has been shown to significantly reduce blood serum levels of enzymes such as alanine aminotransferase (ALT), malondialdehyde (MDA), and aspartate aminotransferase (AST) (González et al. 2003). PC also influences thyroxine hormone (T3) by enhancing the level of serum nitrite (Remirez et al. 2002). R-PE extracted from red algae is another attractive option as an antioxidant for use in photodynamic therapy and cancer prevention. Besides PC and PE, APC also plays a significant role in inhibition of enterovirus 71 in host cells (Shih et al. 2003).

Table 8.2 Fluorescence applications of phycobiliproteins (PBPs)

Fluorescence application	References
Fluorochromes	Hulspas et al. (2009)
Low molecular weight fluorescent labels	Waggoner (2006)
DNA or protein probes	Benvin et al. (2007)
Fluorescence resonance energy transfer (FRET)	Sekar and Chandramohan (2008)
Fluorescence-activated cell sorting	Eriksen (2008) and Sekar and Chandramohan (2008)
Fluorescent probes	Sekar and Chandramohan (2008)
Immunoassays	Telford et al. (2001a)
Intracellular labeling applications	Telford et al. (2001b)
Photocurrent generation	Ihssen et al. (2014)
Multicolor fluorescence applications	Sekar and Chandramohan (2008) and Richa et al. (2011)

8.3.2 Fluorescence Agents

The autofluorescence nature of PBPs has indispensable scope in various experimental approaches requiring fluorescent probes. Inside cyanobacterial cells, PBPs act as accessory light-harvesting protein complexes to harvest light energy for photosynthesis. The individual subunits PC, PE, and APC have a highly fluorescent nature after disintegration from the supramolecular structure of the photosystem. Currently, the demand for natural fluorescent proteins is increasing in the market because of their versatile applications (Table 8.2).

Certain specific spectral properties—such as high excitation/emission, a large wavelength shift, a high molar extinction coefficient, and a high florescence quantum yield—are found in PBPs. Therefore, PBPs have advantages over conventional fluorescent agents. Some important methods used in cell biology—such as flow cytofluorimetry, fluorescence microscopy, single-cell analysis, fluorescence-activated cell sorting (FACS), and immunoassays—are frequently used with the involvement of PBPs as the principal fluorescent agents (Sun et al. 2003; Eriksen 2008; Sekar and Chandramohan 2008). Many chromophores are associated with $\alpha_6\beta_6$ hexamers of every subunit of PBPs that enhance the molar coefficients (Thoren et al. 2006). The same extinction coefficients are reduced and the fluorescence response declines after denaturation of PBPs (Fukui et al. 2004; Kupka and Scheer 2008). The yellow fluorescence nature of PE ($\alpha_6\beta_6$ hexamers) is widely used in fluorescent probes in industry (Glazer 1994). The quantum yield of PE was recorded as being in the range of 82–98% as compared with other subunits (Oi et al. 1982). However, structural configuration of APC and PC has shown reductions in the quantum yield of about 68% and 50%, respectively (Oi et al. 1982). The remarkable nature of PE fluorescence is due to the far shifting of emissions (580 nm) after excitation at 488 nm. PE is being used as a second best colored fluorescent product for fluorescein- labeled antibodies in molecular biology (Sekar and Chandramohan

2008). The growth of these fluorescent products may result in diagnostic tools for cancer, human immunodeficiency virus (HIV), and other deadly diseases in the near future. Recently, multicolor fluorescence analyses have been achieved by association of donor PE and acceptor cyanine dyes Cy5/Cy7 (indodicarbocyanine/indotricarbocyanine). The same combination has an ability to shift the emission wavelength from red to the deep red spectrum, which enhances color analysis of any cells (Waggoner 2006). The wide florescence nature of PE has also been used for development of Affymetrix chips for DNA microarrays. The integrated combination of PE–streptavidin shows a strong signal in biotin-labeled DNA/protein probes for array detection (Benvin et al. 2007). Conjugate association of PBPs with protein A, immunoglobulins, avidin, and biotin have been reported for indispensable dual-color single-cell analysis by FACS (Oi et al. 1982). Interestingly, a fluorochrome (PBXL-3L) has been designated for visualization and immune detection of immunoglobulins in cells (Telford et al. 2001a). Cryptomonad-derived PBPs have low molecular weights and have been evaluated for significant utilization in flow cytometry for extra- and intracellular labeling by fluorescent probes (Telford et al. 2001b). Similarly, genetically stabilized streptavidin–PC has specific molecular domains for development of a site-specific molecular probe (Cai et al. 2001). In the current scenario, most industries are primarily focusing on development of fluorescence labeling reagents for sensitive detection of the fluorescence of single cells. PBPs are multi-chromophore fluorescence proteins that permit sensitive detection of biomolecules in flowcytometry industry. The color properties of PC have been utilized for ecological monitoring (Sode et al. 1991; Simis et al. 2005) and detection of toxic cyanobacteria (Izydorczyk et al. 2005) in the natural environment.

8.4 Cosmetics

Modern research and development have expanded the commercial applications of PBPs (Richa et al. 2011). Moreover, PBPs have played extensive roles in sustainable development of cosmetic products (Fig. 8.4).

PC and PE are marketed in Japan as coloring agents for cosmetics (Prasanna et al. 2007). In Japan, food regulations mandate the application of food colors without any addition of synthetic preservatives. Lina-Blue is a colored commercial dye made by DIC Corporation, which is effectively utilized for natural cosmetics and food products. In the present scenario, a number of companies are designing novel and eco-friendly PC-based products including bath powder, body lotion, whitening cream, soap, eyeliner, lipstick, etc. The PC-/PE-based cosmetics market may achieve high economic growth in the near future for sustainable production of healthy cosmetics.

Fig. 8.4 Cosmetics applications of phycobiliproteins (PBPs)

8.5 Conclusion

In the past few decades, PBPs have been extensively studied to resolve the mystery behind the glowing colors of PC, PE, and APC. Now, incorporation of PBPs into the development of food and nutraceuticals is dominant in certain countries. Certain PBPs have unique properties such as high quantum yields and constancy over a broad range of pH, which enables their use for production of more stable fluorescence products for promising approaches in biotechnological, pharmacological, and biomedical science. There is a need to search for more potential organisms for sustained shelf life in different thermal, light, aqueous, pH, and alcohol-based environments for generation of simple and cost-effective food and pharmaceutical products for human welfare.

References

Arasaki S, Arasaki T (1983) Low calorie, high nutrition vegetables from the sea to help you look and feel better. Japan Publications, Tokyo

Batista AP, Raymundo A, Sousa I, Empis J (2006) Rheological characterization of coloured oil-in-water food emulsions with lutein and phycocyanin added to the oil and aqueous phases. Food Hydrocoll 20:44–52

Benedetti S, Benvenuti F, Pagliarani S et al (2004) Antioxidant properties of a novel phycocyanin extract from the blue–green alga *Aphanizomenon flos-aquae*. Life Sci 75:2353–2362

Benvin AL, Creeger Y, Fisher GW, Ballou B, Waggoner AS, Armitage BA (2007) Fluorescent DNA nanotags: supramolecular fluorescent labels based on intercalating dye arrays assembled on nanostructured DNA templates. J Am Chem Soc 129(7):2025–2034

Bermejo P, Piñero E, Villar ÁM (2008) Iron-chelating ability and antioxidant properties of phycocyanin isolated from a protean extract of *Spirulina platensis*. Food Chem 110:436–445

Bhaskar SU, Gopalaswamy G, Raghu R (2005) A simple method for efficient extraction and purification of c-phycocyanin from *Spirulina platensis* Geitler. Indian J Exp Biol 43:277–279

Bhat VB, Madyastha KM (2000) C-phycocyanin: a potent peroxyl radial scavenger in vivo and in vitro. Biochem Biophys Res Commun 275(1):20–25

Brul S, Coote P (1999) Preservative agents in foods: mode of action and microbial resistance mechanisms. Int J Food Microbiol 50:1–17

Cai YA, Murphy JT, Wedemaye GJ, Glazer AN (2001) Recombinant phycobiliproteins. Recombinant C-phycocyanins equipped with affinity tags, oligomerization, and biospecific recognition domains. Anal Biochem 290:186–204

Chaiklahan R, Chirasuwana N, Bunnag B (2012) Stability of phycocyanin extracted from *Spirulina* sp.: influence of temperature, pH and preservatives. Process Biochem 47:659–664

Cherng S-C, Cheng S-N, Tarn A, Chou T-C (2007) Anti-inflammatory activity of c-phycocyanin in lipopolysaccharide-stimulated RAW 264.7 macrophages. Life Sci 81:1431–1435

Cyanotech Corporation (2007) Spirulina Pacifica. Cyanotech. [Online]. http://www.cyanotech.com/spirulina.html

DIC (2009) Private communication. Tokyo: s.n. 2009

Dong C, Wang W (2006) Headspace solid-phase micro extraction applied to the simultaneous determination of sorbic and benzoic acids in beverages. Anal Chim Acta 562:23–29

Dufosse L, Galaup P, Yaron A, Arad SM, Blanc P, Murthy KNC, Ravishankar GA (2005) Microorganisms and microalgae as sources of pigments for food use: a scientific oddity or an industrial reality? Trends Food Sci Technol 16:389–406

Earthrise Nutritionals LLC (2004) Earthrise the company. Earthrise Nutritionals. [Online]. http://www.earthrise.com/company.asp?page=page5.html

Eriksen NT (2008) Production of phycocyanin—a pigment with applications in biology, biotechnology, foods and medicine. Appl Microbiol Biotechnol 80:1–14

Estrada JPN, Bescós PB, Villar del Fresno AM (2001) Antioxidant activity of different fractions of *Spirulina platensis* protean extract I1. Farmaco 56:497–500

Farooq SM, Asokan D, Kalaiselvi P, Sakthivel R, Varalakshmi P (2004) Prophylactic role of phycocyanin: a study of oxalate mediated renal cell injury. Chem Biol Interact 149:1–7

Fukui K, Saito T, Noguchi Y, Kodera Y, Matsushima A, Nishimura H, Inada Y (2004) Relationship between color development and protein conformation in the phycocyanin molecule. Dyes Pigments 63:89–94

Glazer AN (1981) Photosynthetic accessory proteins with bilin prosthetic groups. In: Hatch MD, Boardman NK (eds) The biochemistry of plants. Academic, New York, pp 51–96

Glazer AN (1985) Light harvesting by phycobilisomes. Annu Rev Biophys Biophys Chem 14:47–77

Glazer AN (1994) Phycobiliproteins—a family of valuable, widely used fluorophores. J Appl Phycol 6:105–112

González R, Rodríques S, Romay C, Ancheta O, González A, Armesto J, Remirez D, Merino N (1999) Anti-inflammatory activity of phycocyanin extract in acetic acid-induced colitis in rats. Pharmacol Res 39:55–59

González R, Gonzalez A, Remirez D (2003) Protective effects of phycocyanin on galactosamine—induced hepatitis in rats. Biotecnol Apl 20:107–110

Grossman AR, Schaefer MR, Chiang GG, Collier JL (1993) The phycobilisomes a light harvesting complex responsive to environmental conditions. Microbiol Rev 57:725–749

Grossman A, Bhaya D, Apt K, Kohoe D (1995) Light-harvesting complexes in oxygenic photosynthesis: diversity, control and evolution. Annu Rev Genet 29:231–288

Holdt SL, Kraan S (2011) Bioactive compounds in seaweed: functional food applications and legislation. J Appl Phycol 23:543–597

Houghton JD (1996) Haems and bilins. In: Houghton JD, Hendry GAF (eds) Natural food colorants, 2nd edn. Blackie (Chapman and Hall), Glasgow, pp 157–196

Hulspas R, Dombkowski D, Preffer F, Douglas D, Kildew-Shah B, Gilbert J (2009) Flow cytometry and the stability of phycoerythrin–tandem dye conjugates. Cytometry A 75(11):966–972

Ihssen J, Artur B, Faccio G, Gajda-Schrantz K, Thöny-Meyer L (2014) Light harvesting proteins for solar fuel generation in bioengineered photoelectrochemical cells. Curr Protein Pept Sci 15(4):374–384

Izydorczyk K, Tarczynska M, Jurczak T, Mrowczynski J, Zalewski M (2005) Measurement of phycocyanin fluorescence as an online early warning system for cyanobacteria in reservoir intake water. Environ Toxicol 20:425–430

Jensen GS, Ginsberg DI, Drapeau C (2001) Blue-green algae as an immuno-enhancer and biomodulator. J Am Nutraceut Ass 3:24–30

Jespersen L, Strømdahl LD, Olsen K, Skibsted LH (2005) Heat and light stability of three natural blue colorants for use in confectionery and beverages. Eur Food Res Technol 220:261–266

Kahn M, Varadharaj S, Shobha JC, Naidu MU, Parinandi NL, Kutala VK, Kuppusamy P (2006) C-phycocyanin ameliorates doxorubicin-induced oxidative stress and apoptosis in adult rat cardiomyocytes. J Cardiovasc Pharmacol 47:9–20

Kannaujiya VK, Sinha RP (2015) Impacts of varying light regimes on phycobiliproteins of *Nostoc* sp. HKAR-2 and *Nostoc* sp. HKAR-11 isolated from diverse habitats. Protoplasma 252:1551–1561

Kannaujiya VK, Sinha RP (2016) Thermokinetic stability of phycocyanin and phycoerythrin in food grade preservatives. J Appl Phycol 28:1063–1070

Kumar CS, Ganesan P, Suresh PV, Bhaskar N (2008) Seaweeds as a source of nutritionally beneficial compounds-a review. J Food Sci Technol 45:1–13

Kupka M, Scheer H (2008) Unfolding of C-phycocyanin followed by loss of non-covalent chromophore–protein interactions. 1. Equilibrium experiments. Biochim Biophys Acta 1777:94–103

Lee Y-K (1997) Commercial production of microalgae in the Asia–Pacific rim. J Appl Phycol 9:403–411

Leema JT, Kirubagaran R, Vinithkumar NV, Dheenan PS, Karthikayulu S (2010) High value pigment production from *Arthrospira (Spirulina) platensis* cultured in seawater. Bioresour Technol 101:9221–9227

Lino CM, Pena A (2010) Occurrence of caffeine, saccharin, benzoic acid and sorbic acid in soft drinks and nectars in Portugal and subsequent exposure assessment. Food Chem 121:503–508

Liu Y, Xu L, Cheng N et al (2000) Inhibitory effect of phycocyanin from *Spirulina platensis* on the growth of human leukemia k562 cells. J Appl Phycol 12:125–130

Mishra SK, Shrivastav A, Pancha I, Jain D, Mishra S (2010) Effect of preservatives for food grade C-phycoerythrin, isolated from marine cyanobacteria *Pseudanabaena* sp. Int J Biol Macromol 47:597–602

Mishra SK, Shrivastav A, Maurya RR, Patidar SK, Haldar S, Mishra S (2012) Effect of light quality on the C-phycoerythrin production in marine cyanobacteria *Pseudanabaena* sp. isolated from Gujarat coast, India. Protein Expr Purif 81:5–10

Nagaoka S, Shimizu K, Kaneko H et al (2005) A novel protein C-phycocyanin plays a crucial role in the hypocholesterolemic action of *Spirulina platensis* concentrate in rats. J Nutr 135:2425–2430

Oi VT, Glazer AN, Stryer L (1982) Fluorescent phycobiliprotein conjugates for analyses of cells and molecules. J Cell Biol 93:981–986

Prasanna R, Sood A, Suresh A, Kaushik BD (2007) Potentials and applications of algal pigments in biology and industry. Acta Bot Hung 49:131–156

Remirez D, Fernandez V, Tapia G (2002) Influence of C-phycocyanin on hepatocellular parameters related to liver oxidative stress and Kupffer cell functioning. Inflamm Res 51:351–356

Richa KVK, Kesheri M, Singh G, Sinha RP (2011) Biotechnological potentials of phycobiliproteins. Int J Pharm Bio Sci 2:446–454

Rimbau V, Camins A, Romay C, González R, Pallàs M (1999) Protective effects of C-phycocyanin against kainic acid-induced neuronal damage in rat hippocampus. Neurosci Lett 276:75–78

Riss J, Décordé K, Sutra T, Delage M, Baccou J-C, Jouy N, Brune J-P, Oréal M, Cristol J-P, Rouanet J-M (2007) Phycobiliprotein C-phycocyanin from *Spirulina platensis* is powerfully responsible for reducing oxidative stress and NADPH oxidase expression induced by an atherogenic diet in hamsters. J Agric Food Chem 55:7962–7967

Roman BR, Alvarez-Pez JM, Fernandez A et al (2002) Recovery of pure B-phycoerythrin from the microalga *Porphyridium cruentum*. J Biotechnol 93:73–85

Romay C, Armesto J, Remirez D, González R, Ledon N, García I (1998) Antioxidant and anti-inflammatory properties of C-phycocyanin from bluegreen algae. Inflamm Res 47(1):36–41

Romay CH, Gonzalez R, Ledon N et al (2003) C-phycocyanin a biliprotein with antioxidant, ant-inflammatory and neuroprotective effects. Curr Protein Pept Sci 4:207–216

Roy KR, Arunasree KM, Reddy NP, Dheeraj B, Reddy GV, Reddanna P (2007) Alteration of mitochondrial membrane potential by *Spirulina platensis* C-phycocyanin induces apoptosis in the doxorubicin-resistant human hepatocellular-carcinoma cell line HepG2. Biotechnol Appl Biochem 47:159–167

Sekar S, Chandramohan M (2008) Phycobiliproteins as a commodity: trends in applied research, patents and commercialization. J Appl Physiol 20:113–136

Shih SR, Tsai KN, Li YS et al (2003) Inhibition of enterovirus 71-induced apoptosis by allophycocyanin isolated from a blue green alga *Spirulina platensis*. J Med Virol 70:119–125

Simis SGH, Peters SWM, Gons HJ (2005) Remote sensing of the cyanobacterial pigment phycocyanin in turbid inland water. Limnol Oceanogr 50:237–245

Singh S, Kate BN, Banerjee UC (2005) Bioactive compounds from cyanobacteria and microalgae: an overview. Crit Rev Biotechnol 25:73–95

Sode KJ, Horikoshi K, Takeyama J, Nakamura N, Matsunga T (1991) On line monitoring of marine cyanobacterial cultivation based on phycocyanin fluorescence. J Biotechnol 21:209–218

Spolaore P, Joannis-Cassan C, Duran E, Isambet A (2006) Commercial applications of microalgae. J Biosci Bioeng 101(2):87–96

Subhashini J, Mahipal VK, Reddy MC, Reddy MM, Rachamallu A, Reddanna P (2004) Molecular mechanisms in C-phycocyanin induced apoptosis in human chronic myeloid leukemia cell line-K562. Biochem Pharmacol 68:453–462

Sun L, Wang S, Chen L, Gong X (2003) Promising fluorescent probes from phycobiliproteins. IEEE J Sel Top Quantum Electron 9:177–188

Telford WG, Moss MW, Moreseman JP et al (2001a) Cryptomonad algal phycobiliproteins as fluorochromes for extracellular and intracellular antigen detection by flow cytometry. Cytometry 44:16–23

Telford WG, Moss MW, Moreseman JP et al (2001b) Cyanobacterial stabilized phycobilisomes as fluorochromes for extracellular antigen detection by flow cytometry. J Immunol Methods 254:13–30

Thoren KL, Connell KB, Robinson TE, Shellhamer DD, Tammaro MS, Gindt YM (2006) The free energy of dissociation of oligomeric structure in phycocyanin is not linear with denaturant. Biochemistry 45:12050–12059

Waggoner A (2006) Fluorescent labels for proteomics and genomics. Curr Opin Chem Biol 10:62–66

Role in Therapeutic Sciences 9

9.1 Introduction

Phycobiliproteins (PBPs) have widespread biotechnological applications including nutraceuticals, pharmaceuticals, food industry, cosmetics, agriculture, bioremediation, aquaculture, biofuels, and bioenergy (Abed et al. 2009; Pandey 2010; Richa et al. 2011). PC is water-soluble, brilliantly blue colored, and most widely used as natural fluorescent and pharmaceutical agents in various biotechnological applications. However, PE and APC are mainly used for fluorescent, cosmetics, and certain other biotechnological applications. Recent analysis of patent informatics about PBPs shows that 297 patents filed in global patent databases; 51 patents on production; 30 patents on food, clinical, cosmetics, and pharmaceutical applications; and 216 patents on fluorescent applications (Sekar and Chandramohan 2008; Eriksen 2008). Although numerous applications have been found on commercial production of PBPs from various cyanobacteria species, however, certain cyanobacterium like *Spirulina* sp. (*Arthrospira*) is much exploited for large-scale production (Chen et al. 1996; Chaneva et al. 2007). To enhance the diversity of PBP-based bioproduct, biotechnological companies of leading country have followed a wide range of extraction and purification technology to obtain higher product quality. Thus, modern research and development about novel structural configuration biliproteins have expanded the further application of PBPs in many therapeutic sciences (Table 9.1). PC also acts as natural photosensitizer that is used for light-oriented therapy for tumor cells called as photodynamic therapy (Zhang et al. 1995). PBP-based drugs are still in progress for use in various clinical applications. In this chapter, we have focused on therapeutic significance and probable mechanism of action of PBPs in various human diseases.

V.K. Kannaujiya et al., *Phycobiliproteins: Recent Developments and Future Applications*, https://doi.org/10.1007/978-981-10-6460-9_9

Table 9.1 Biomedical applications of phycobiliproteins

Biomedical application	References
Anti-inflammatory	González et al. (1999), Benedetti et al. (2004), and Riss et al. (2007)
Antioxidant	Miranda et al. (1998), Romay et al. (1999) and Bermejo et al. (2008)
Liver protecting	Kutay et al. (1995), Qureshi et al. (1996), and Remirez et al. (2002a, b)
Neuroprotective agent	Marín-Prida et al. (2012) and Mitra et al. (2015)
Antimutagenic	Chamorro et al. (1996)
Immunity enhancer	Cian et al. (2012)
Antiviral	Ayehunie et al. (1998)
Antiallergic	Kim et al. (1998)
Lipase inhibitor and serum lipid-reducing agent	Nagaoka et al. (2005)
Blood-lipid-lowering effects	Iwata et al. (1990)
Atherosclerosis	Riss et al. (2007) and Gupta et al. (2011)
Prevents oxalic acid-induced kidney stone formation, inhibit renal toxicity	Zheng et al. (2013) and Farooq et al. (2014)
Reduce cardiotoxicity	Kahn et al. (2006)
Enzyme inhibition represses nitric oxide synthase	Remirez et al. (2002a, b), Cherng et al. (2007), and Chen et al. (2012)
Antiplatelet aggregation	Chiu et al. (2006)
Anticancerogenic	Schwartz et al. (1988), Chen and Zhang (1995), Liu et al. (2000, 2016), Subhashini et al. (2004), and Roy et al. (2007)
Cataract	Kumari et al. (2013, 2015)

9.2 Therapeutic Behavior of Various Diseases

9.2.1 Anticancer

PC is extensively used as blue colorant for various food and cosmetics due to high neutraceutical value, protein nature, high-stroke shift, and strong fluorescence in ultraviolet spectrum. PC has a potent anticancer effect in variety of cancer cell types including lung cancer (Li et al. 2015), breast cancer (Li et al. 2010), bone marrow cancer (Gardeva et al. 2014), and colon cancer (Cai et al. 1995). Although PC purified from *Spirulina platensis* shows distinct anticancer activity, heavy molecular weight and complex structure may hinder determination of molecular mechanism. Prior to introduction of PC as a drug in organism, PC was disintegrated into low molecular weight peptide by the process of enzymatic hydrolysis and column chromatography and subjected to tumor inhibitory response on HeLa and 293T tumor cells that shows significant better response as compared to non-disintegrated PC (Wang et al. 2012). The component of α and β subunits of PC also influenced the growth response of lung cancer cell line SPC-A-1 (Sun et al. 2010; Zhang et al.

2010). In comparison to α subunits, β subunits show distinguished effects on growth of tumor. The β subunit of PC is very active and binds to tubulin proteins and metabolic enzyme such as glyceraldehyde-3-phosphate dehydrogenase (GAPDH) which accelerated caspase-3 and caspase-9 factor, therefore cell cycle arrested in the G0/G1 phase leading to apoptosis and inhibition of growth of tumor (Liu et al. 2016).

9.2.2 Anti-inflammatory

Anti-inflammatory responses of PC have been well described (Liu et al. 2016), and it was first reported by Remirez et al. (2002a). PC is used as nutritional proteins for treatment to osteoarthritis which reduce various inflammatory cytokines, interleukin-6 (IL-6), TNF-, NO, MMP-3, and sulfated glycosaminoglycans (Martinez et al. 2015). PC can able to inhibit cyclooxygenase-2 activity and promotes leukotriene B4 formation in mouse ear inflammation test (Romay et al. 1999; Reddy et al. 2000). Nitric oxide (NO) that is synthesized by the catalytic regulation of nitric oxide synthase (NOS) plays an important role in regulation of physiological and pathological inflammation response of cell (Moncada et al. 1991). There are two NOS isoforms that have been found such as Ca^{2+}/calmodulin-dependent NOS (constitutive) called cNOS and another Ca^{2+}/calmodulin-independent NOS (inducible) called as iNOS (Forstermann et al. 1991). The presence of high amount of NO stimulated pro-inflammatory cytokines and generated free radicals and lipopolysaccharide (LPS) that is a critical mediator for pathogenesis-related inflammatory diseases (Vane et al. 1994; Szabo and Thiemermann 1995). Shih et al. (2009) found that PC can inhibit excessive level of NO and PGE2 by downregulation of iNOS and COX-2 expression which reduce the formation of TNF- and infiltration of neutrophils to inflammation sites. Remirez et al. (2002b) had shown inhibitory effects on allergic inflammatory response mediated by inhibition of histamine release from mast cells.

9.2.3 Antioxidant

Recently, PC is exploited in a variety of biotechnology and pharmacological products. In addition, PC is also approved as potent antioxidant, neuroprotective, and hepatoprotective role in various diseases (Romay et al. 2003). PC has ability to scavenge hydroxyl, alkoxyl, peroxyl, peroxynitrite ($ONOO^-$), and hypochlorous acid (HOCl) and plays potent role as antioxidant in organism (Bermejo et al. 2008). Free radicals are very dangerous and have the ability to penetrate cells and cause many disorders including inflammation, atherosclerosis, cancer, reperfusion injury, and other diseases (Kehrer 1993; Lynch 1998; Gupta et al. 2011). PC has a potent neuroprotective agent. They show protective effects on tributyltin chloride-induced neurotoxicity along with N-acetylcysteine (neuroprotective drug) (Mitra et al. 2015). PC is inhibited Ca2+/phosphate-induced mitochondrial damage in rat brain by inhibition of membrane potential, ROS levels, and proapoptotic Cyt c (Marín-Prida et al. 2012). PC is also used to treat neurodegenerative diseases including ischemic stroke, Alzheimer's disease, and Parkinson's disease (Rimbau et al. 1999; Marín-Prida et al. 2012).

PC can significantly reduce liver toxicity and protect liver enzymes like P450, aminopyrine-N-demethylase, and glucose-6-phosphatase (Liu et al. 2016). Thus, PC plays critical role as hepatoprotective to protect the liver enzymes. The antioxidant property of PC also helps in reduction of hepatic brain injury by induction of thioacetamide (Kutay et al. 1995). Recently, it has been shown that PC has effects on Kupffer cell function to reduce phagocytosis pathway in response of oxidative stress-induced production of tumor necrosis factor alpha- (TNF-) and nitric oxide (NO) promoted by hyperthyroid state (Remirez et al. 2002b).

PC can inhibit cisplatin-triggered renal toxicity and oxidative stresses which help in preservation of antioxidative enzyme and mark attenuation of oxidative stress (Fernandez-Rojas et al. 2014). PC can able to prevent cellular damage by increasing oxalic acid-mediated oxidative stress in canine kidney cells (Farooq et al. 2014). Moreover, PC also prevents the diabetic nephropathy by inhibition of NADPH-dependent superoxide production in renal mesangial cells (Zheng et al. 2013).

PC has played a significant role in reduction of cardiovascular disease (CVD) by inhibiting lipid metabolism, mitochondrial damage, and oxidative stress. PC effectively reduces inflammatory damage induced by oxidative stress in atherosclerosis and increases formation of COX-2 which further increases antioxidant enzymes in the body and regulates blood lipid level for proper functioning of the heart (Riss et al. 2007). Thus, PC can inhibit progress of atherosclerosis by promoting expression of CD59 and reduction of blood fat, muscle cell proliferation, and apoptosis of endothelial cell (Li et al. 2013).

Cataract is age-related disorder caused by oxidative stress. PC can adjust antioxidative levels which reduces chance of cataract in rat (Kumari et al. 2013). PC can regulate transcriptional level of lens crystalline, redox, and expression mRNA for apoptotic cascade pathway which play significant role in maintenance of lens transparency (Kumari et al. 2015).

9.3 Molecular View of Therapeutic Science

9.3.1 Gene Expression and Signal Transduction Mechanism

Apoptosis plays an important role in development and homoeostasis of many diseases, including HIV, cancer, pathogenic infection, and AIDS (Roshal et al. 2001; Stuart and Hughes 2002; Li et al. 2006). Apoptosis is a primary mode for cell death recognized by a certain specific characteristic feature in terms of morphological and molecular alteration in cells. Several reports have revealed the presence of a number of genes such as p53 (antioncogene), Fas/FasL (proapoptotic gene), nuclear transcription factor, Bcl-2, and caspase gene, which inhibit the process of apoptosis and development of cancer (Choi et al. 2004; Cao et al. 2008).

Cytochrome c (Cyt c) originated from mitochondria and plays an important role in electron transport for oxidation of biomolecules. Poly-ADP-ribose polymerase

(PARP) is another tumor-inhibiting enzyme that plays a distinguished role in DNA repair and cell apoptosis (Reddy et al. 2003). The deregulation of Fas/FasL, ICAM-1 (intercellular cell-adhesion molecule 1 CD54), and Bcl-2 (B-cell lymphocytic leukemia proto-oncogene 2) is an indication for development of tumors; thereafter oncogene mechanism directs regulation of tumors (Li et al. 2006). The extract of PC from *Spirulina platensis* induced phenomena of apoptotic pathway in macrophage cell line (RAW 264.7) (Reddy et al. 2003) and rat histiocytoma cell line (AK5) (Pardhasaradhi et al. 2003). PC also induced apoptosis in HeLa cells with marked common morphological changes such as cell shrinkage, membrane blebs, condensation of nuclear and cytoplasm, endolytic cleavage nucleic acids, generation of apoptotic bodies, and micronuclei approved by electron microscopy (Li et al. 2006). Primarily, PC is triggered to arrest the cell cycle at G0/G1 phase in human tumor cell, and cell proliferation becomes block due to inhibition of DNA synthesis (Basha et al. 2008; Yong et al. 2001). Caspase is an aspartic acid protease activity having cysteine residue responsible for selective cutting of proteins to facilitate cell apoptosis (Ying et al. 2015). The activity of caspases 2, 3, 4, 6, 8, 9, and 10 was functional in PC-treated HeLa cells that indicate that caspase-dependent pathway may facilitate apoptosis (Li et al. 2006). PC was also found to inhibit pancreatic cell line (PANC-1) by inactivation of BCR-ABL signaling, downstream PI3K/Akt pathway1 (Tantirapan and Suwanwong 2014), and upregulation of caspase 3 and caspase 8 activities (Wang et al. 2007). Further, PC has been demonstrated for apoptosis inducer in tumor cells by accelerating ROS production and downregulating the expression of Bcl-2 (Pardhasaradhi et al. 2003) and inducing cytochrome c found in cytosol of mitochondria (Subhashini et al. 2004). It is well known that NF-κB may regulate the apoptotic pathway in both positive and negative manners (Papa et al. 2004; Javelaud and BesancËon 2001; Schlottmann et al. 2008; Hwang et al. 2010; Trocoli and Djavaheri-Mergny 2011). PC is also suppressing activity for Akt/mTOR pathway which facilitates for induction of NF-κB in response to DNA damage, whereas mTOR deficiency may induce phosphorylation of nuclear factor that enhanced the activity of NF-κB. PC induced autophagy by suppressing the Akt/mTOR pathway, whose activation counteracts the pro-survival effects of NF-κB activation on gene expression in response to DNA damage and cytokines (Ghosh et al. 2006). The mTOR deficiency or inactivation increases phosphorylation and nuclear translocation of nuclear factor NF-κB, which results in enhanced NF-κB activation (Guo et al. 2013). Moreover, increasing key transcription initiator factor Beclin 1 enhances cell autophagy by NF-κB pathway (Copetti et al. 2009). Recently, an important signaling mechanism has been described that PC is able to induce apoptotic and autophagic cancer cell through homeostatic phenomena of MAPK, NF-κB, and Akt/mTOR signaling pathways (Liao et al. 2016). There are several signaling molecules which regulated in the presence of PC pigment (Fig. 9.1).

COX-2 is an enzyme-accelerated tumor formation (Huang et al. 2015) and progression (Esbona et al. 2016), tumor angiogenesis (Waskewich et al. 2002; Tabriz et al. 2016). The reducing or blocking tumor development process is primarily regulated by prostaglandin E2 (PGE2). Current development may increase nondependent pathway including inhibition of Bcl-2 gene (Chen and Ren 2007) and vascular

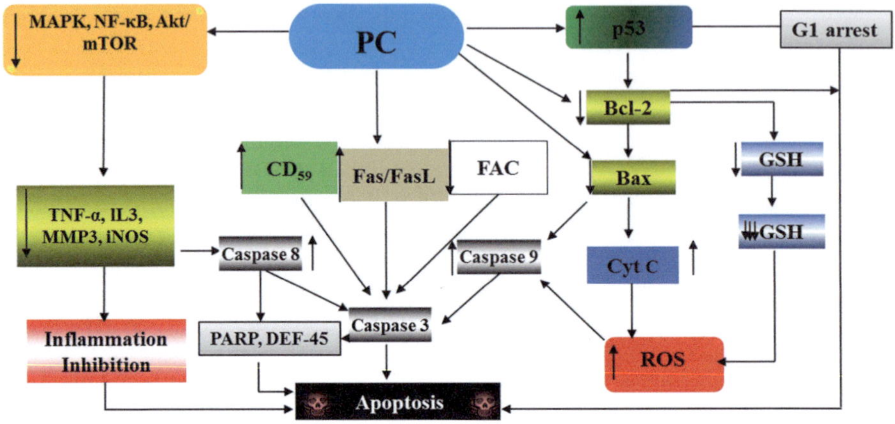

Fig. 9.1 Mechanism of regulation of PC-induced genes and proteins (Modified from Liu et al. 2016). Up arrow mark, upregulation; down arrow mark, downregulation

endothelial growth factor (VEGF) (von Rahden et al. 2005; Tao et al. 2006), for regulation of COX-2 enzyme, matrix metalloprotease (MMP), and urokinase-type plasminogen.

9.3.2 Combinatorial Response of PC with Other Drugs

9.3.2.1 Combination of PC and Betaine

The combined effects of PC and betaine on p38 MAPK pathway and cell cycle G_2/M phase are well recognized (Yi and Kim 2012; Lee 2015; Wu et al. 2011). Moreover, combined effects of PC and betaine were further approved on growth of A549 tumor by analyses of growth inhibition, cell viability, and proapoptotic pathway (Bingula et al. 2016). In therapeutic science, PC increased efficacy of chemotherapeutic anticancer drug like cisplatin and carboplatin in selective manner (Bhattacharyya and Sharma 2016) which indicates PC-based drug may be involved in the treatment of cancer.

9.3.2.2 Combination of PC and Piroxicam

Piroxicam is oxicam class drug used to relieve pain and inflammation in rheumatoid arthritis, osteoarthritis, and primary dysmenorrhea. This is a kind of nonsteroidal, anti-inflammatory drug that inhibits the production of prostaglandins which triggered pain, tenderness, stiffness, and swelling during arthritis. The combination of piroxicam and PC has shown positive effects on rat colon carcinogenesis by more reduction of tumor number and size of tumor as compared to single piroxicam drug. Thus, piroxicam combined with PC has accelerated the effectiveness by reducing cytotoxicity, high drug delivery response, and tumor progression (Saini et al. 2012, 2013; Saini and Sanyal 2012, 2014a, b, 2015).

9.3.2.3 Combination of PC and Tretinoin

Tretinoin also called all-*trans* retinoic acid (ATRA) is commonly used for treatment of acne and acute leukemia to inhibit tumor development process. It is applied on skin as skin ointment or cream for treatment of cancer. The combination of PC and tretinoin has shown better response on A549 lung cancer cell line as compared to individual tretinoin (Li et al. 2015, 2016). The effective dose of tretinoin required is very less in amount in combination with PC without any toxic side effects.

9.3.2.4 Combination of PC and Topotecan

Topotecan (TPT) is a water-soluble chemotherapeutic agent and found in trade name of Hycamtin in the market. Topotecan inhibits topoisomerase I activity and is used for treatment of almost all kinds of solid tumor including lung cancer, ovarian cancer, and prostate cancer. The combination of TPT and PC shows positive effects on cancer cell line. The combination of 10% TPT and PC has shown more effects on prostate cancer as compared to more than 10% TPT individually by increasing caspase-3 and caspase-9 expression (Gantar et al. 2012).

9.3.2.5 Combination of PC and Doxorubicin

Doxorubicin (DOX) is an anthracycline-containing antitumor and antibiotic drug that is used in chemotherapy for treatment of several types of cancer. Doxorubicin commonly worked on inhibition of DNA/RNA synthesis. The combination of DOX and PC has shown better effects on hepatocellular carcinoma cell line (HepG2) as compared to individual DOX (Roy et al. 2007; Nishanth et al. 2010). Thus, PC may enhance the efficacy of any anticancer drug.

9.4 Emerging Technology

Photodynamic therapy (PDT) is an emerging technology and noninvasive method to treat deadly untreatable disease like cancer, psoriasis, and bacterial and viral infectious diseases (Bharathiraja et al. 2016). This therapy has many advantages over conventional chemotherapy, localized surgery, and overcoming drug resistance mechanisms like antibiotics. PDT is a process where intense solar light and photosensitizers used for production of free radicals lead to cell death of cancer (Wainwright 2004). Many photosensitizers have been characterized such as chlorophyll-derived protoporphyrin IX and chlorin e6, but due to its hydrophobic nature, it aggregates easily in physiological solution that reduces the efficacy of PDT mechanism (Ormond and Freeman 2013). PC has many advantages over chlorophyll-derived photosensitizer due to its high water solubility nature, nontoxicity, and immune-modulating properties (Muthulakshmi et al. 2012; Chen et al. 2012, 2014a, b). PC is easily metabolized in normal cells as compared to cancerous cells; therefore, it could be used in PDT for eradication of cancer without any harm to normal cells (Bharathiraja et al. 2016). Recently, nonthermal irradiation like low-level laser therapy (LLLT) has been developed for reducing pain and inflammation during PDT mechanism (Fig. 9.2). This emerging technology has been used for treatment of various kinds of

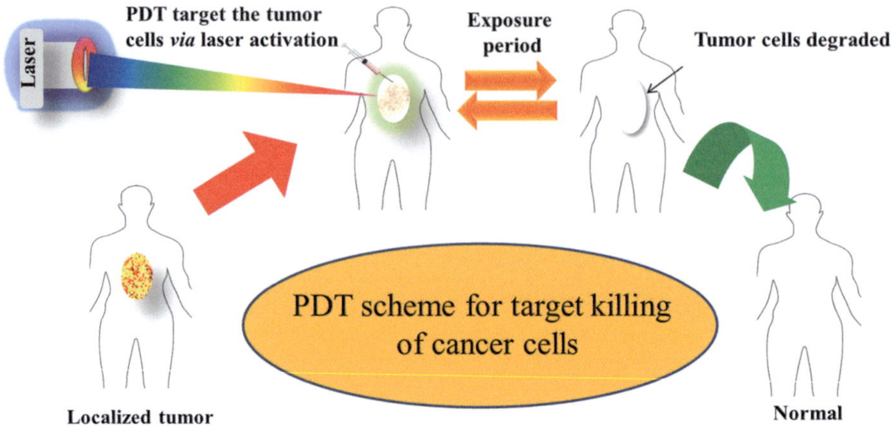

Fig. 9.2 Proposed models for photodynamic therapy of localized tumor in human

diseases like myocardial infraction, tumor eradication, traumatic brain disorders, and other cancer-oriented diseases (Avci et al. 2013). The cancer cell line such as HeLa cell tumor and breast cancer MCF-7 cells have shown tumor cell death and immune enhancement after PC-based PDT therapy (Li et al. 2010, 2011).

9.5 Conclusion

Several kinds of diseases arise to increase dependency of allopathic medicine which affects human health constantly. Cancer is one of the most serious diseases and major threats to human health. Although, some satisfactory treatment has been found for certain diseases but still several challenges exists due to change in our lifestyle. PC is widely studied for treatment of oxidative-induced disease; cardiovascular disease; hepatoprotective, neural disease; and cancer-related major disease. The involvement of PC-based PDT scheme is more advantageous for oxidative-induced treatment of cancer and bacterial and viral disease. PC plays strong role in immune enhancement that induces self-healing ability and high recovery rate after chemotherapy surgery and radiotherapy. In the context of rapid research and development for discovery, PC-based new formulation may now glimmer for treatment of several kinds of untreatable diseases. Therefore, PC and other PBP components such as PE and APC offer great food products, nutraceuticals, and pharmaceutical and biomedical products for human welfare.

References

Abed RMM, Dobretsov S, Sudesh K (2009) Applications of cyanobacteria in biotechnology. J Appl Microbiol 106:1–12

Avci P, Gupta A, Sadasivam M, Vecchio D, Pam Z, Pam N, Hamblin MR (2013) Low-level laser (light) therapy (LLLT) in skin: stimulating, healing, restoring. Semin Cutan Med Surg 32:41–52

Ayehunie S, Belay A, Baba TW, Ruprecht RM (1998) Inhibition of HIV-1 replication by an aqueous extract of *Spirulina platensis (Arthrospira platensis)*. J Acquir Immune Defic Syndr Hum Retrovirol 18:7–12

Basha OM, Hafez RA, El-Ayouty YM, Mahrous KF, Bareedy MH, Salama AM (2008) C-phycocyanin inhibits cell proliferation and may induce apoptosis in human HepG2 cells. Egypt J Immunol 15:161–167

Benedetti S, Benvenutti F, Pagliarani S, Francogli S, Scoglio S, Canestrari F (2004) Antioxidant properties of a novel phycocyanin extract from the blue-green alga *Aphanizomenon flos-aquae*. Life Sci 75:2353–2362

Bermejo P, Piñero E, Villar AM (2008) Iron-chelating ability and antioxidant properties of phycocyanin isolated from a protean extract of *Spirulina platensis*. Food Chem 110:436–445

Bharathiraja S, Seo H, Manivasagan P, Moorthy MS, Park S, Oh J (2016) In vitro photodynamic effect of phycocyanin against breast cancer cells. Molecules 21(11):1470

Bhattacharyya S, Sharma V (2016) C-phycocyanin modulates the cytotoxicity of platinum based anticancer drugs in lung cancer cell lines. Int J Sci Res 5:2

Bingula R, Dupuis C, Pichon C, Berthon J-Y, Filaire M, Pigeon L, Filaire E (2016) Study of the effects of betaine and/or C-phycocyanin on the growth of lung cancer A549 cells in vitro and in vivo. J Oncol 2016:8162952

Cai X-h, He L, Jiang-Jialun J, Xu X, Zheng S (1995) The experimental study of application of phycocyanin in cancer laser therapy. Chin Mar Drug 1:15–18

Cao B, Lei ZY, Chen H et al (2008) YB-13, a novel synthetic microtubule inhibitor, induces apoptosis of HeLa cells and its mechanism. Chin Pharmacol Bull 1:123–127

Chamorro G, Salazar M, Favila L, Bourges H (1996) Pharmacology and toxicology of *Spirulina* alga. Rev Investig Clin 48:389–399

Chaneva G, Furnadzhieva S, Minkova K, Lukavsky J (2007) Effect of light and temperature on the cyanobacterium *Arthronema africanum*- a prospective phycobiliprotein-producing strain. J Appl Phycol 19:537–544

Chen JJ, Ren H (2007) Research progress of anti-tumor effect of COX-2 selective inhibitors. China Pharm 18:1508–1509

Chen F, Zhang Q (1995) Inhibitive effects of *spirulina* on aberrant crypts in colon induced by dimethylhydrazine. Zhonghua Yu Fang Yi Xue Za Zhi 29:13–17

Chen F, Zhang Y, Guo S (1996) Growth and phycocyanin formation of *Spirulina platensis* in photoheterotrophic culture. Biotechnol Lett 18:603–608

Chen JC, Liu KS, Yang TJ et al (2012) *Spirulina* and C-phycocyanin reduce cytotoxicity and inflammation-related genes expression of microglial cells. Nutr Neurosci 15:252–256

Chen H-W, Yang T-S, Chen M-J, Chang Y-C, Eugene I, Wang C, Ho C-L, Lai Y-J, Yu C-C, Chou J-C (2014a) Purification and immunomodulating activity of C-phycocyanin from *Spirulina platensis* cultured using power plant flue gas. Process Biochem 49:1337–1344

Chen EP, Markosyan N, Connolly E et al (2014b) Myeloidcell COX-2 deletion reduces mammary tumor growth through enhanced cytotoxic T-lymphocyte function. Carcinogenesis 35:1788–1797

Cherng S-C, Cheng S-N, Tarn A, Chou T-C (2007) Anti-inflammatory activity of c-phycocyanin in lipopolysaccharide-stimulated RAW 264.7 macrophages. Life Sci 81:1431–1435

Chiu H-F, Yang S-P, Kuo Y-L, Lai Y-S, Chou T-C (2006) Mechanisms involved in the antiplatelet effect of C-phycocyanin. Br J Nutr 95:435–440

Choi S-E, Sohn S, Cho J-W, Shin E-A, Song P-S, Kang Y (2004) 9-hydroxypheophorbide -induced apoptotic death of MCF-7 breast cancer cells is mediated by c-Jun N-terminal kinase activation. J Photochem Photobiol B Biol 73:101–107

Cian RE, Lopez-Posadas RL, Drago SR, de Medina FS, Martinez-Augustin O (2012) Immunomodulatory properties of the protein fraction from *Phorphyra columbina*. J Agric Food Chem 60:8146–8154

Copetti T, Bertoli C, Dalla E, Demarchi F, Schneider C (2009) p65/RelA modulates BECN1 transcription and autophagy. Mol Cell Biol 29:2594–2608

Eriksen NT (2008) Production of phycocyanin-a pigment with applications in biology, biotechnology, foods and medicine. Appl Microbiol Biotechnol 80:1–14

Esbona K, Inman D, Saha S et al (2016) COX-2 modulates mammary tumor progression in response to collagen density. Breast Cancer Res 18:35

Farooq SM, Boppana NB, Devarajan A, Sekaran SD, Shankar EM (2014) C-phycocyanin confers protection against oxalate mediated oxidative stress and mitochondrial dysfunctions in MDCK cells. PLoS One 9:e103361

Fernandez-Rojas B, Medina-Campos ON, Hernandez-Pando R, Negrette-Guzman M, Huerta-Yepez S, Pedraza-Chaverri J (2014) C-phycocyanin prevents cisplatin-induced nephrotoxicity through inhibition of oxidative stress. Food Funct 5:480–490

Forstermann U, Schmidt HHW, Pollock JS, Sheng H, Mitchell JA, Warner TD, Nakane M, Murand R (1991) Isoforms of nitric oxide synthase. Characterization and purification from different cell types. Biochem Pharmacol 42:1849–1857

Gantar M, Dhandayuthapani S, Rathinavelu A (2012) Phycocyanin induces apoptosis and enhances the effect of topotecan on prostate cell line LNCaP. J Med Food 15:1091–1095

Gardeva E, Toshkova E, Yossifova L, Minkova K, Ivanova N, Gigova L (2014) Antitumor activity of C-phycocyanin from *Arthronema africanum* (Cyanophyceae). Braz Arch Biol Technol 57:675–684

Ghosh S et al (2006) Essential role of tuberous sclerosis genes TSC1 and TSC2 in NF-κB activation and cell survival. Cancer Cell 10:215–226

González R, Rodríques S, Romay C, Ancheta O, González A, Armesto J, Remirez D, Merino N (1999) Anti-inflammatory activity of phycocyanin extract in acetic acid-induced colitis in rats. Pharmacol Res 39:55–59

Guo F et al (2013) MTOR regulates DNA damage response through NF-κB-mediated FANCD2 pathway in hematopoietic cells. Leukemia 27:2040–2046

Gupta M, Dwivedi UN, Khandelwal S (2011) C-phycocyanin: an effective protective agent against thymic atrophy by tributyltin. Toxicol Lett 204:2–11

Huang J, Zhang D, Xie F, Lin D (2015) The potential role of COX-2 in cancer stem cell-mediated canine mammary tumor initiation: an immunohistochemical study. J Vet Sci 16:225–231

Hwang J, Lee H-J, Lee W-H, Suk K (2010) NF-κB as a common signaling pathway in ganglioside-induced autophagic cell death and activation of astrocytes. J Neuroimmunol 226:66–72

Iwata K, Inayama T, Kato T (1990) Effects of *Spirulina platensis* on plasma lipoprotein lipase activity in fructose-induced hyperlipidemic rats. J Nutr Sci Vitaminol 36:165–171

Javelaud D, BesancËon F (2001) NF-kB activation results in rapid inactivation of JNK in TNFa-treated Ewing sarcoma cells: a mechanism for the anti-apoptotic effect of NF-kB. Oncogene 20:4365–4372

Kahn M, Varadharaj S, Shobha JC, Naidu MU, Parinandi NL, Kutala VK, Kuppusamy P (2006) C-phycocyanin ameliorates doxorubicin-induced oxidative stress and apoptosis in adult rat cardiomyocytes. J Cardiovasc Pharmacol 47:9–20

Kehrer JP (1993) Free radicals as mediators of tissue injury and disease. Crit Rev Toxicol 23:21–48

Kim HM, Lee EH, Cho HH, Moon YH (1998) Inhibitory effect of mast cell-mediated immediate-type allergic reactions in rats by *Spirulina*. Biochem Pharmacol 55:1071–1076

Kumari RP, Sivakumar J, Thankappan B, Anbarasu K (2013) C-phycocyanin modulates selenite-induced cataractogenesis in rats. Biol Trace Elem Res 151:59–67

Kumari RP, Ramkumar S, Thankappan B et al (2015) Transcriptional regulation of crystallin, redox, and apoptotic genes by C-phycocyanin in the selenite-induced cataractogenic rat model. Mol Vis 21:26–39

Kutay U, Ahnert-Hilger G, Hartmann E, Wiedenmann B, Rapoport TA (1995) Transport route for synaptobrevin *via* a novel pathway of insertion into the endoplasmic reticulum membrane. EMBO J 14:217–223

Lee I (2015) Betaine is a positive regulator of mitochondrial respiration. Biochem Biophys Res Commun 456:621–625

Li B, Gao M-H, Zhang X-C, Chu X-M (2006) Molecular immune mechanism of C-phycocyanin from *Spirulina platensis* induces apoptosis in HeLa cells in vitro. Biotechnol Appl Biochem 43:155–164

Li B, Chu X, Gao M, Li W (2010) Apoptotic mechanism of MCF-7 breast cells in vivo and in vitro induced by photodynamic therapy with C-phycocyanin. Acta Biochim Biophys Sin 42:80–89

Li B, Chu XM, Gao MH (2011) Treatment of hela tumor in mice with c-phycocyanin mediated photodynamic therapy and its immune mechanism underlying apoptosis. Chin J Laser Med Surg 20:1–6

Li B, Chu X-M, Xu Y-J, Yang F, Lv C-Y, Nie S-M (2013) CD59 underlines the antiatherosclerotic effects of C-phycocyanin on mice. BioMed Res Int 2013:729413

Li B, Gao MH, Chu XM et al (2015) The synergistic antitumor effects of all-trans retinoic acid and C-phycocyanin on the lung cancer A549 cells in vitro and in vivo. Eur J Pharmacol 749:107–114

Li B, Gao MH, Lv CY, Yang P, Yin QF (2016) Study of the synergistic effects of all-transretinoic acid and C-phycocyanin on the growth and apoptosis of A549 cells. Eur J Cancer Prev 25:97–101

Liao G, Gao B, Gao Y, Yang X, Cheng X, Ou Y (2016) Phycocyanin inhibits tumorigenic potential of pancreatic cancer cells: role of apoptosis and autophagy. Sci Rep 6:34564

Liu Q, Huang Y, Zhang R, Cai T, Yu C (2016) Medical application of Spirulina platensis derived C-phycocyanin. Evid Based Complement Alternat Med 2016:7803846

Lynch MA (1998) Age-related impairment in long-term potentiation in hippocampus: a role for the cytokine, interleukin-1? Prog Neurobiol 56:571–589

Marín-Prida J, Pentón-Rol G, Rodrigues FP et al (2012) C-phycocyanin protects SH-SY5Y cells from oxidative injury, rat retina from transient ischemia and rat brain mitochondria from Ca2+/phosphate-induced impairment. Brain Res Bull 89:159–167

Martinez SE, Chen Y, Ho EA, Martinez SA, Davies NM (2015) Pharmacological effects of a C-phycocyanin-based multicomponent neutraceutical in an in-vitro canine chondrocyte model of osteoarthritis. Can J Vet Res 79:241–249

Miranda MS, Cintra RG, Barros SB, Mancini Filho J (1998) Antioxidant activity of the microalga *Spirulina maxima*. Braz J Med Biol Res 31:1075–1079

Mitra S, Siddiqui WA, Khandelwal S (2015) C-phycocyanin protects against acute tributyltin chloride neurotoxicity by modulating glial cell activity along with its anti-oxidant and anti-inflammatory property: a comparative efficacy evaluation with N-acetyl cysteine in adult rat brain. Chem Biol Interact 238:138–150

Moncada S, Palmer RMJ, Higgs EA (1991) Nitric oxide: physiology, pathophysiology, and pharmacology. Pharmacol Rev 43:109–142

Muthulakshmi M, Saranya A, Sudha M, Selvakumar G (2012) Extraction, partial purification, and antibacterial activity of phycocyanin from *Spirulina* isolated from fresh water body against various human pathogens. J Algal Biomass Util 3:7–11

Nagaoka S, Shimizu K, Kaneko H et al (2005) A novel protein C-phycocyanin plays a crucial role in the hypocholesterolemic action of *Spirulina platensis* concentrate in rats. J Nutr 135:2425–2430

Nishanth RP, Ramakrishna BS, Jyotsna RG et al (2010) Cphycocyanin inhibits MDR1 through reactive oxygen species and cyclooxygenase-2 mediated pathways in human hepatocellular carcinoma cell line. Eur J Pharmacol 649:74–83

Ormond AB, Freeman HS (2013) Dye sensitizers for photodynamic therapy. Materials 6:817–840

Pandey VD (2010) Bioremediation of heavy metals by microalgae. In: Das MK (ed) Algal biotechnology. Daya Publishing House, Delhi, pp 287–296

Papa S, Zazzeroni F, Pham CG, Bubici C, Franzoso G (2004) Linking JNK signaling to NF-κB: a key to survival. J Cell Sci 117:5197–5208

Pardhasaradhi BVV, Mubarak AA, Leela KA, Reddanna P, Ashok K (2003) Phycocyanin mediated apoptosis in AK-5 tumor cells involves down regulation of Bcl-2 and generation of ROS. Mol Cancer Ther 2:1165–1170

Qureshi MA, Garlich JD, Kidd MT (1996) Dietary *Spirulina platensis* enhances humoral and cell-mediated immune functions in chickens. Immunopharmacol Immunotoxicol 18:465–476

von Rahden BHA, Stein HJ, Pühringer F et al (2005) Coexpression of cyclooxygenases (COX-1, COX-2) and vascular endothelial growth factors (VEGF-A, VEGF-C) in esophageal adenocarcinoma. Cancer Res 65:5038–5044

Reddy CM, Bhat VB, Kiranmai G, Reddy MN, Reddanna P, Madyastha KM (2000) Selective inhibition of cyclooxygenase-2 by C-phycocyanin, a biliprotein from *Spirulina platensis*. Biochem Biophys Res Commun 277:599–603

Reddy MC, Subhashini J, Mahipal SV, Bhat VB, Srinivas Reddy P, Kiranmai G, Madyastha KM, Reddanna P (2003) C-phycocyanin, a selective cyclooxygenase-2 inhibitor, induces apoptosis in lipopolysaccharide-stimulated RAW 264.7 macrophages. Biochem Biophys Res Commun 304:385–392

Remirez D, Fernandez V, Tapia G, Gonzalez R, Videla LA (2002a) Influence of C-phycocyanin on hepatocellular parameters related to liver oxidative stress and Kupffer cell functioning. Inflamm Res 51:351–356

Remirez D, Ledón N, González R (2002b) Role of histamine in the inhibitory effects of phycocyanin in experimental models of allergic inflammatory response. Mediat Inflamm 11:81–85

Richa, Kannaujiya VK, Kesheri M, Singh G, Sinha RP (2011) Biotechnological potentials of phycobiliproteins. Int J Pharma Bio Sci 2:B446–454

Rimbau V, Camins A, Romay C, Gonzalez R, Pallás M (1999) Protective effects of C-phycocyanin against kainic acid induced neuronal damage in rat hippocampus. Neurosci Lett 276:75–78

Riss J, Décordé K, Sutra T et al (2007) Phycobiliprotein C-phycocyanin from *Spirulina platensis* is powerfully responsible for reducing oxidative stress and NADPH oxidase expression induced by an atherogenic diet in hamsters. J Agric Food Chem 55:7962–7967

Romay C, Ledon N, Gonzalez R (1999) Phycocyanin extract reduces leukotriene B4 levels in arachidonic acid-induced mouse ear inflammation test. J Pharm Pharmacol 51:641–642

Romay CH, Gonzalez R, Ledon N et al (2003) C-phycocyanin a biliprotein with antioxidant, antinflammatory and neuroprotective effects. Curr Protein Pept Sci 4:207–216

Roshal M, Zhu Y, Planelles V (2001) Apoptosis in AIDS. Apoptosis 6:103–116

Roy KR, Arunasree KM, Reddy NP, Dheeraj B, Reddy GV, Reddanna P (2007) Alteration of mitochondrial membrane potential by *Spirulina platensis* C-phycocyanin induces apoptosis in the doxorubicin resistant human hepato cellular carcinoma cell line HepG2. Biotechnol Appl Biochem 47:159–167

Saini MK, Sanyal SN (2012) PTEN regulates apoptotic cell death through PI3-K/Akt/GSK3 signaling pathway in DMH induced early colon carcinogenesis in rat. Exp Mol Pathol 93:135–146

Saini MK, Sanyal SN (2014a) Targeting angiogenic pathway for chemoprevention of experimental colon cancer using C-phycocyanin as cyclooxygenase-2 inhibitor. Biochem Cell Biol 92:206–218

Saini MK, Sanyal SN (2014b) Piroxicam and c-phycocyanin prevent colon carcinogenesis by inhibition of membrane fluidity and canonical Wnt/-catenin signaling while up-regulating ligand dependent transcription factor PPAR. Biomed Pharmacother 68:537–550

Saini MK, Sanyal SN (2015) Cell cycle regulation and apoptotic cell death in experimental colon carcinogenesis: intervening with cyclooxygenase-2 inhibitors. Nutr Cancer Int J 67:620–636

Saini MK, Sanyal SN, Vaiphei K (2012) Piroxicam and C-phycocyanin mediated apoptosis in 1,2-dimethylhydrazine dihydrochloride induced colon carcinogenesis: exploring the mitochondrial pathway. Nutr Cancer 64:409–418

Saini MK, Vaish V, Sanyal SN (2013) Role of cytokines and Jak3/Stat3 signaling in the 1,2-dimethylhydrazine dihydrochloride-induced rat model of colon carcinogenesis: early target in the anticancer strategy. Eur J Cancer Prev 22:215–228

Schlottmann S, Buback F, Stahl B, Meierhenrich R, Walter P et al (2008) Prolonged classical NF-κB activation prevents autophagy upon E. coli stimulation in vitro: a potential resolving mechanism of inflammation. Mediat Inflamm 2008:725854

Schwartz J, Shklar G, Reid S, Trickler D (1988) Prevention of experimental oral cancer by extracts of *Spirulina-Dunaliella* algae. Nutr Cancer 11:127–134

Sekar S, Chandramohan M (2008) Phycobiliproteins as a commodity: trends in applied research, patents and commercialization. J Appl Physiol 20:113–136

Shih C-M, Cheng S-N, Wong C-S, Kuo Y-L, Chou T-C (2009) Antiinflammatory and antihyperalgesic activity of Cphycocyanin. Anesth Analg 108:1303–1310

Stuart L, Hughes J (2002) Apoptosis and autoimmunity. Nephrol Dial Transplant 17:697–700

Subhashini J, Mahipal SV, Reddy MC et al (2004) Molecular mechanisms in C-phycocyanin induced apoptosis in human chronic myeloid leukemia cell line-K562. Biochem Pharmacol 68:453–462

Sun GY, Liang H, Xu QY (2010) Study on antitumor activity of phycocyanin and its antioxidant function. Prog Modern Biomed 10:243–245

Szabo C, Thiemermann C (1995) Regulation of the expression of the inducible isoform of nitric oxide synthase. Adv Pharmacol 34:113–153

Tabriz HM, Mirzaalizadeh M, Gooran S, Niki F, Jabri M (2016) COX-2 expression in renal cell carcinoma and correlations with tumor grade, stage and patient prognosis. Asian Pac J Cancer Prev 17:535–538

Tantirapan P, Suwanwong Y (2014) Anti-proliferative effects of C-phycocyanin on a human leukemic cell line and induction of apoptosis via the PI3K/AKT pathway. J Chem Pharm Res 6:1295–1301

Tao K-X, Zhang N, Wang G-B, Wu X-B (2006) Effects of meloxicam on vascular endothelial growth factor and angiopoietin-2 expression in colon carcinoma cell line HT-29. World Chin J Dig 14:1277–1282

Trocoli A, Djavaheri-Mergny M (2011) The complex interplay between autophagy and NF-κB signaling pathways in cancer cells. Am J Cancer Res 1:629

Vane JR, Mitchell JA, Appleton I, Tomlinson A, Bishop-Bailey D, Croxtall J, Willoughby DA (1994) Inducible isoforms of cyclooxygenase and nitric oxide synthase in inflammation. Proc Nat Acad Sci USA 91:2046–2050

Wainwright M (2004) Photodynamic therapy-from dyestuffs to high-tech clinical practice. Rev Prog Color Relat Top 34:95–109

Wang H, Liu Y, Gao X, Carter CL, Liu Z-R (2007) The recombinant β subunit of C-phycocyanin inhibits cell proliferation and induces apoptosis. Cancer Lett 247:150–158

Wang XQ, Deng W, Yang JF, Mao YC, Shi ZM (2012) Isolation and purification and cytotoxic activity of enzymatic hydrolysis peptide of phycocyanin. Food Sci 1:136–140

Waskewich C, Blumenthal RD, Li H, Stein R, Goldenberg DM, Burton J (2002) Celecoxib exhibits the greatest potency amongst cyclooxygenase (COX) inhibitors for growth inhibition of COX-2-negative hematopoietic and epithelial cell lines. Cancer Res 62:2029–2033

Wu L-C, Lin Y-Y, Yang S-Y, Weng Y-T, Tsai Y-T (2011) Antimelanogenic effect of c-phycocyanin through modulation of tyrosinase expression by upregulation of ERK and downregulation of p38 MAPK signaling pathways. J Biomed Sci 18:74

Yi E-Y, Kim Y-J (2012) Betaine inhibits in vitro and in vivo angiogenesis through suppression of the NF-κB and Akt signaling pathways. Int J Oncol 41:879–1885

Ying J, Pan RW, Wang MF et al (2015) Effects of phycocyanin on apoptosis of human laryngeal cancer HEP-2 cells. Chin J Pathophysiol 7:1189–1196

Yong W, Qian F, Qian KX, Dong Q (2001) Anticancer activity of phycocyanin. J Zhejiang Uni (Eng Sci) 35:672–675

Zhang CW, Yin ZM, Ou YPK (1995) Exploitation and utilization of phycobiliprotein. Chin J Mar Drugs 3:52–53

Zhang X, Li JY, Gong XG (2010) Isolation of C-PC subunits from *Spirulina platensis* and inhibitory effect on SPC-A-1 cell line. J Zhej Uni 37:319–323

Zheng J, Inoguchi T, Sasaki S et al (2013) Phycocyanin and phycocyanobilin from *Spirulina platensis* protect against diabetic nephropathy by inhibiting oxidative stress. Am J Physiol-Regul Integr Comp Physiol 304:110–120

Future Development and Challenges 10

Cyanobacteria are pioneer photosynthetic oxygen-evolving ancient prokaryotes and most abundant nitrogen-fixing microorganisms commonly found in rice paddy fields, marine, hot spring, and Arctic and Antarctic habitats (Häder et al. 2015). They play an extensive role in global photosynthetic biomass production and movement of nutrient cycle. Cyanobacteria are dependent on solar energy for photosynthesis and nitrogen fixation. The accessory light-harvesting antenna complexes (phycobiliproteins) are major component in cyanobacteria and red algae that exhibit incredible colors and play a vital role in the harvesting of light energy from the sun. Phycobiliproteins (PBPs) are brilliantly colored water-soluble and autofluorescence that are present in cyanobacteria, red algae, and cryptomonads. Nucleotides like GC play a distinguished role in nucleotide variation of phycocyanin (PC), phycoerythrin (PE), and allophycocyanin (APC) by broad range of mutation and translational selection (Lynch et al. 2008; Yu et al. 2012). Nucleotide variation on the third position (GC3) has more specific codon expression as compared to the first and second nucleotide position (Kannaujiya et al. 2014). Thus nucleotide regulation of PC/PE and APC is still to be discovered for efficient production of more stable PBPs from cyanobacteria. Bilin lyase enzymes are playing an important role in attachment of chromophore to PC and PE bilin precursor. There are four types of lyases that have been described: (1) E/F-type lyases, (2) S(U)-type lyases, (3) T-type lyases, and (4) Y/Z-type lyase (Biswas et al. 2010). The strains of *Escherichia coli* are serving as a model organism for investigation of biosynthetic pathways of PBPs. Several models have been proposed for biosynthesis of PBPs. However, homologous expressions of gene expression of PBPs are still to be discovered for productive biosynthesis and functional properties. The current advancement in gene manipulation and recombinant synthesis of PBPs may substantially add to future development of research. The molecular mechanism related to structure and light-induced signaling pathways in photosystem is well described. However, the structural alterations in morphology of cell are not well understood. The secondary messengers have amplifying environmental signals received from primary messengers (light) in cyanobacteria (Agostoni and Montgomery 2014). Understanding the biosynthetic pathway of

V.K. Kannaujiya et al., *Phycobiliproteins: Recent Developments and Future Applications*, https://doi.org/10.1007/978-981-10-6460-9_10

PBPs may fulfill the limitation and add in global economy. Knowledge about PBPs may serve as a key point to improve engineering and biotechnological processes for critical adaptation and biosynthesis against changing environment. Recently, possible pathway of energy transfer has been remodeled (Johnson et al. 2014; Kannaujiya et al. 2016). However, role of amino acid and structural integrity of PBPs in energy storage and transfer is still to be understood. In the current scenario, there is a tremendous loss of fossil fuels due to a rapid demand of energy. Therefore, an utmost need is to search renewable source of energy to reduce future energy crisis. The mechanism of energy harvesting in cyanobacteria is crucial for development of green energy production. Now, current challenges are to focus on PBPs-Chl integrated system for development of green energy device. Apart from naïve structural information, PBPs have ability to regulate their internal composition and property in response to various abiotic environmental signals like nutrient availability, light intensity, and temperature (Prassana et al. 2004). The abiotic stress plays an important role in sustainable production of PBPs from cyanobacteria. Moreover, continuous increase in ultraviolet radiation may cause alterations in biologically important molecules like proteins, nucleic acids, and other relevant molecules. Thus, it affects a number of vital physiological and biochemical functions in living organisms (Sinha et al. 2002). The high intensity of light may reduce the fluorescence nature of PBPs (Rastogi et al. 2015 ; Kannaujiya and Sinha 2017a, b). Concurrent abiotic stress shows rapid degradation in PBP composition, which severally affects biotechnological and production for commercial utilization industrial. Thus, close adjustment of abiotic stress of environment may enhance the productivity of PBPs from any cyanobacterial species. There is a need to search more cyanobacteria to sustain shelf life in thermal, light, aqueous, pH, and alcoholic environment for generation of simple and cost-effective food and pharmaceutical products for human welfare. Recently, exclusive demand of PBPs continuously increases in reference to selection of productive strain (Larkum et al. 2012) and genetic engineering of metabolic pathway (Georgianna and Mayfield 2012). The various methodologies of cultivation types of PC/PE/APC production have been reported (Eriksen 2008; Kuddus et al. 2013). Recently, several tubular closed bioreactors have been designed for large-scale microalgae cultivation with automatic operation and highly controlled system for continuous production. However, large-scale production of value-added compounds in closed bioreactor is limited due to high investment of cost and energy (Borowitzka and Moheimani 2013; Tredici et al. 2015). The development of algal film bioreactor is a better approach for production of high-value compounds. Alternatively, recombinant genetic engineering is another technique for production of certain compounds which could be utilized in large-scale production of particular compounds. Therefore, the biomass industries are utmost needed to overcome certain limitations such as cost of production, energy utilization, and imbalance of high-value food products for proper utilization of biomass for biotechnology industry. In the current scenario, natural food color market is tremendously increasing worldwide. The data of market analysis of 2016 shows that food color business was reached US$ 1.3 billion with 6.8% growth rate annually. The current growth rate is expected to reach US$1.77 billion up to 2021. To increase sustainable use and quality of PBP-based product, there is need to enhance purity index. Currently, several techniques have been used for extraction and purification of PBPs. Indeed, none of common method is developed which is applicable for extraction and purification of PBPs from any cyanobacterial sources with high yield and high recovery. Thus, an urgent need is to develop a commercial technique for effective production and more utilization of PBPs to fulfill demand in mankind. The future depends on the growth and development of bioprocess engineering for manufacturing as well as improvement of commercial products. Modern food industry leads to an increase in the production of cheaper, healthier, and more convenient food products without any harm to the human body. Therefore, consumer's demand for more natural food products, having health benefits, has increased over the years. PBPs are one of the most promising food products that are being used efficiently by people as additive and marketed as food and cosmetic colorant in Japan, China, India, and other European and Asian countries. Apart from nutritional value of PBPs, it stimulates the immune defense system and possesses antioxidant, anti-inflammatory, antiviral, anticancer, and cholesterol-lowering effects. PC is more stable than indigo and gardenia and emits a bright blue fluorescent color in jelly gum, soft candies, fermented milk products, ice creams, soft drinks, desserts, sweet cake decoration, milk shakes, and cosmetics (Richa et al. 2011). Sekar and Chandramohan (2008) counted status of patents as such 55 for production of PBPs, 236 for fluorescence application, and 30 for clinical importance.

PC has a potent anticancer effect in variety of cancer cell types including lung cancer (Li et al. 2015), breast cancer (Li et al. 2010), bone marrow cancer (Gardeva et al. 2014), and colon cancer (Cai et al. 1995). The β subunit of PC is very active and binds to tubulin proteins and metabolic enzyme such as glyceraldehyde-3-phosphate dehydrogenase (GAPDH) which accelerates caspase-3 and caspase-9 factors, therefore cell cycle arrested in the G0/G1 phase leading to apoptosis and inhibition of growth of tumor (Liu et al. 2016). PC plays a significant role in reduction of cardiovascular disease (CVD) by inhibition of lipid metabolism, mitochondrial damage, and oxidative stress. Recently, an important signaling mechanism has been described for PC-induced apoptotic and autophagy cancer cell through homeostatic phenomena of MAPK, NF-κB, and Akt/mTOR signaling pathways (Liao et al. 2016). There are several signaling molecules which regulated in the presence of PC pigment. Photodynamic therapy (PDT) is an emerging technology and noninvasive method to treat deadly untreatable disease like cancer, psoriasis, and bacterial and viral infectious diseases (Bharathiraja et al. 2016). This therapy has many advantages over conventional chemotherapy, localized surgery, and overcoming drug resistance mechanisms like antibiotics. The involvement of PC-based PDT scheme is more advantageous for oxidative-induced treatment of cancer, bacterial and viral diseases. PC plays a strong role in immune enhancement that induces self-healing ability and high recovery rate after chemotherapy, surgery and radiotherapy.

References

Agostoni M, Montgomery BL (2014) Survival strategies in the aquatic and terrestrial world: the impact of second messengers on cyanobacterial processes. Life 4:745–769

Bharathiraja S, Seo H, Manivasagan P, Moorthy MS, Park S, Oh J (2016) In vitro photodynamic effect of phycocyanin against breast cancer cells. Molecules 21(11):1470

Biswas A, Vasquez YM, Dragomani TM, Kronfel ML, Williams SR, Alvey RM et al (2010) Biosynthesis of cyanobacterial phycobiliproteins in *Escherichia coli*: chromophorylation efficiency and specificity of all bilin lyases from *Synechococcus* sp. strain PCC 7002. Appl Environ Microbiol 76:2729–2739

Borowitzka MA, Moheimani NR (2013) Algae for biofuels and energy. Developments in applied phycology. Springer, Dordrecht, p 285

Cai X-h, He L, Jiang-Jialun J, Xu X, Zheng S (1995) The experimental study of application of phycocyanin in cancer laser therapy. Chinese Marine Drug 1:15–18

Eriksen NT (2008) Production of phycocyanin-a pigment with applications in biology, biotechnology, foods and medicine. Appl Microbiol Biotechnol 80:1–14

Gardeva E, Toshkova E, Yossifova L, Minkova K, Ivanova N, Gigova L (2014) Antitumor activity of C-phycocyanin from *Arthronema africanum* (Cyanophyceae). Braz Arch Biol Technol 57:675–684

Georgianna DR, Mayfield SP (2012) Exploiting diversity and synthetic biology for the production of algal biofuels. Nature 488:329–335

Häder D-P, Williamson CE, Wängberg S, Rautio M, Rose KC, Gao K, Helbling EW, Sinha RP, Worrest R (2015) Effects of UV radiation on aquatic ecosystems and interactions with other environmental factors. Photochem Photobiol Sci 14:108–126

Johnson EM, Kumar K, Das D (2014) Physicochemical parameters optimization, and purification of phycobiliproteins from the isolated *Nostoc* sp. Bioresour Technol 166:541–547

Kannaujiya VK, Sinha RP (2017a) Detection of free thiols and fluorescence response of phycoerythrin chromophore after ultraviolet-B radiation stress. J Fluoresc 27:561–567

Kannaujiya VK, Sinha RP (2017b) Impacts of diurnal variation of ultraviolet-B and photosynthetically active radiation on phycobiliproteins of the hot-spring cyanobacterium *Nostoc* sp. strain HKAR-2. Protoplasma 254:423–433

Kannaujiya VK, Rastogi RP, Sinha RP (2014) GC constituents and relative codon expressed amino acid composition in cyanobacterial phycobiliproteins. Gene 546:162–171

Kannaujiya VK, Rahman A, Adinath SS, Sinha RP (2016) Structural and functional dynamics of tyrosine amino acid in phycocyanin of hot-spring cyanobacteria: a possible pathway for internal energy transfer. Gene Rep 5:83–91

Kuddus M, Singh P, Thomas G, Al-Hazimi A (2013) Recent developments in production and biotechnological applications of c-phycocyanin. Bio Med Res Int 742859

Larkum AWD et al (2012) Selection, breeding and engineering of microalgae for bioenergy and biofuel production. Trends Biotechnol 30:198–205

Li B, Chu X, Gao M, Li W (2010) Apoptotic mechanism of MCF-7 breast cells in vivo and in vitro induced by photodynamic therapy with C-phycocyanin. Acta Biochim Biophys Sin 42:80–89

Li B, Gao MH, Chu XM, Teng L, Lv CY, Yang P, Yin QF (2015) The synergistic antitumor effects of all-trans retinoic acid and C-phycocyanin on the lung cancer A549 cells in vitro and in vivo. Eur J Pharmacol 749:107–114

Liao G, Gao B, Gao Y, Yang X, Cheng X, Ou Y (2016) Phycocyanin inhibits tumorigenic potential of pancreatic cancer cells: role of apoptosis and autophagy. Sci Rep 6:34564

Liu Q, Huang Y, Zhang R, Cai T, Yu C (2016) Medical application of *Spirulina platensis* derived C-phycocyanin. Evid Based Complement Alternat Med 2016:7803846

Lynch M, Sung W, Morris K, Coffey N, Landry CR, Dopman EB, Dickinson WJ, Okamoto K, Kulkarni S, Hartl DL, Thomas WK (2008) A genome wide view of the spectrum of spontaneous mutations in yeast. Proc Natl Acad Sci U S A 105:9272–9277

Prassana R, Pabby A, Saxena S, Singh PK (2004) Modulation of pigment profiles of *Calothrix elenkenii* in response to environmental changes. J Plant Physiol 161:1125–1132

Rastogi RP, Sonani RR, Madamwar D (2015) Effects of PAR and UV radiation on the structural and functional integrity of phycocyanin, phycoerythrin and allophycocyanin isolated from the marine cyanobacterium *Lyngbya* sp. A09DM. Photochem Photobiol 91:837–844

Richa, Kannaujiya VK, Kesheri M, Singh G, Sinha RP (2011) Biotechnological potentials of phycobiliproteins. Int J Pharma Bio Sci 2:446–454

Sekar S, Chandramohan M (2008) Phycobiliproteins as a commodity: trends in applied research, patents and commercialization. J Appl Physiol 20:113–136

Sinha RP, Richter P, Faddoul J, Braun M, Häder D-P (2002) Effects of UV and visible light on cyanobacteria at the cellular level. Photochem Photobiol Sci 1:553–559

Tredici M, Bassi N, Prussi M, Biondi N, Rodolfi L, Chini Zittelli G, Sampietro G (2015) Energy balance of algae biomass production in a 1-ha "Green Wall Panel" plant: how to produce algae biomass in a closed reactor achieving a high Net Energy Ratio. Appl Energy 154:1103–1111

Yu T, Li J, Yang Y, Qi L, Chen B, Zhao F, Bao Q, Wu J (2012) Codon usage patterns and adaptive evolution of marine unicellular cyanobacteria *Synechococcus* and *Prochlorococcus*. Mol Phylogenet Evol 62:206–213

MIX
Papier aus verantwortungsvollen Quellen
Paper from responsible sources
FSC® C105338

If you have any concerns about our products,
you can contact us on
ProductSafety@springernature.com

In case Publisher is established outside the EU,
the EU authorized representative is:
Springer Nature Customer Service Center GmbH
Europaplatz 3, 69115 Heidelberg, Germany

Printed by Libri Plureos GmbH
in Hamburg, Germany